COLGATE UNIVERSITY LIBRARY

Mass Extinction

Ashraf M. T. Elewa

Mass Extinction

Ashraf M. T. Elewa
Minia University
Fac. Science
Dept. Geology
Minia 61519
Egypt
aelewa@link.net

ISBN: 978-3-642-09504-7   e-ISBN: 978-3-540-75916-4

© 2010 Springer Verlag Berlin Heidelberg

This work is subject to copyright. All rights are reserved, whether the whole or part of the material is concerned, specifically the rights of translation, reprinting, reuse of illustrations, recitation, broadcasting, reproduction on microfilm or in any other way, and storage in data banks. Duplication of this publication or parts thereof is permitted only under the provisions of the German Copyright Law of September 9, 1965, in its current version, and permission for use must always be obtained from Springer. Violations are liable for prosecution under the German Copyright Law.

The use of general descriptive names, registered names, trademarks, etc. in this publication does not imply, even in the absence of a specific statement, that such names are exempt from the relevant protective laws and regulations and therefore free for general use.

Cover design: deblik, Berlin

Printed on acid-free paper

9 8 7 6 5 4 3 2 1

springer.com

Dedication

This book is dedicated to **my mother**

In fact, she is one of the great women who could give an excellent example of how eastern women can be effective, like western women, in developing their societies

Foreword

P. David Polly

Department of Geological Sciences, Indiana University, Bloomington, IN 47405, USA, pdpolly@indiana.edu

Only 200 years ago, extinction was a radical new idea. Fossils were known, but their identity as the remains of species that no longer lived on the face of the Earth was not yet firmly established in the scientific world. Arguments that these organic-looking objects from the rocks were merely bizarre mineralizations or that they were the remains of species still living in unexplored regions of the world had dominated 18th Century interpretations of fossils. But the settling of North America and other colonial expeditions by Europeans were quickly making the world smaller. In 1796 Cuvier painstakingly demonstrated that the anatomy of the mastodon skeleton from Big Bone Lick in Kentucky could not possibly belong to a modern elephant, unlike the mammoth fossils found in Europe, which are so similar to the living African Elephant that many found plausible the explanation that they were bones of animals used by the Roman army. Any doubt that Cuvier's mastodon still lived in the wilds of the western North American interior was crushed ten years later when the Lewis and Clark expedition failed to find any sign of them.

That fossils were the remains of extinct organisms revolutionized scientific understanding of the Earth and its history. The advent of stratigraphic geology in the early decades of the 19th Century brought the realization that there were sometimes major hiatuses in the continuity of life, when entire faunas and floras were replaced by others in the succession of rocks. Some of these hiatuses defined major boundaries in the geological timescale: the break between Permian and Jurassic periods became the dividing line between the Paleozoic and Mesozoic, and the turnover at the end of the Cretaceous became the division of Mesozoic from Cenozoic. Not only were many once living species now extinct, but they often perished together *en masse*. Quantitative work in the 1970s by Raup, Sepkoski, and others again revolutionized understanding of extinctions. Extinctions were periodic and massive, some of them like the end-Permian extinction wiping out 82% of the genera living at the time. Thanks to environmental geochemistry and other lines of evidence, we now understand that the Earth's biota is complexly interlinked with cycles of oxygen, carbon, erosion, tectonics, global temperature, sea level, and ecology. Living organisms are both affected by geoclimatic changes and help drive them.

This new book edited by Prof. Ashraf Elewa presents a combination of reviews and original papers on several major extinctions: the Late Ordovician, the Late Devonian, the end Permian, the end Triassic, the end Cretaceous, and the Late Pleistocene extinctions. Elewa provides reviews of each of these, which are interspersed by original papers by international experts on each one, including Mikael Calner, Curtis Congreve, Ahmed Dakrory, Julien Louys, Spencer Lucas, Lawrence Tanner and Panos Petrakis, in addition to Elewa himself. The book concludes, appropriately, with papers on the current extinction that is being driving by the peopling of the Earth and the rapid environmental changes we have wrought.

Table of Contents

1 Mass Extinction - a general view 1
   Ashraf M. T. Elewa
     References 4

2 Late Ordovician mass extinction 5
   Ashraf M. T. Elewa
     References 7

3 The End Ordovician; an ice age in the middle of a greenhouse 9
   Curtis R. Congreve
     3.1 Introduction 9
     3.2 Early research: The discovery of the glacial period 10
     3.3 Fast or slow: The changing face of the Ordovician glaciation 12
     3.4 Trilobite extinction and larval form.. 15
     3.5 Future work... 17
     3.6 Acknowledgements 17
     References 18

4 Silurian global events – at the tipping point of climate change 21
   Mikael Calner
     4.1 Introduction 21

4.2 The Silurian marine scene 24
　　4.3 Discovery of the Silurian global events.. 25
　　　　4.3.1 Early discoveries... 27
　　　　4.3.2 The theory of Silurian oceanic events 28
　　　　4.3.3 The increased use of stable isotope stratigraphy 29
　　4.4 The Early Silurian Ireviken Event 31
　　　　4.4.1 Stratigraphic position.. 31
　　　　4.4.2 Groups affected.. 33
　　　　4.4.3 Stable isotopes 34
　　　　4.4.4 Sedimentary changes and sea-level 35
　　4.5 The Middle Silurian Mulde Event 35
　　　　4.5.1 Stratigraphic position.. 36
　　　　4.5.2 Groups affected.. 37
　　　　4.5.3 Stable isotopes 38
　　　　4.5.4 Sedimentary changes and sea-level 38
　　4.6 The Late Silurian Lau Event.. 39
　　　　4.6.1 Stratigraphic position.. 40
　　　　4.6.2 Groups affected.. 40
　　　　4.6.3 Stable isotopes 41
　　　　4.6.4 Sedimentary changes and sea-level 42
　　4.7 Structure of the Silurian global events.. 43
　　　　4.7.1 The Silurian events and carbonate platforms.................... 43
　　　　4.7.2 Temporal and spatial development and biodiversity........... 45
　　　　4.7.3 Carbon isotope stratigraphy.. 47
　　　　4.7.4 Sea-level change and sedimentary facies 48
　　　　4.7.5 Taxonomic vs ecologic events... 49
　　4.8 Summary... 49
　　4.9 Acknowledgement 50
　　References 50

5 Late Devonian mass extinction 59
　　Ashraf M. T. Elewa
　　References 60

6 Late Permian mass extinction... 61
　　Ashraf M. T. Elewa
　　Reference... 62

7 Late Triassic mass extinction.. 63
　　Ashraf M. T. Elewa
　　References 64

8 Reexamination of the end-Triassic mass extinction 65
Spencer G. Lucas and Lawrence H. Tanner
8.1 Introduction 65
8.2 Late Triassic Timescale 66
8.3 Some methodological issues 66
8.4 Extinctions of taxonomic groups 71
    8.4.1 Radiolarians 71
    8.4.2 Marine bivalves 73
    8.4.3 Ammonoids 74
    8.4.4 Reef builders 75
    8.4.5 Conodonts 78
    8.4.6 Land plants 80
    8.4.7 Tetrapods 82
8.5 Ecological severity 85
8.6 Causes of the TJB extinctions 86
8.7 Late Triassic extinction events 88
8.8 Acknowledgments 90
References 90

9 Cenomanian/Turonian mass extinction of macroinvertebrates in the context of Paleoecology; A case study from North Wadi Qena, Eastern Desert, Egypt 103
Ahmed Awad Abdelhady
9.1 Introduction 103
9.2 Geographical and geologic setting 105
9.3 Material and methods 106
9.4 Correlation of diversity and preservation 107
9.5 Biostratigraphy 108
9.6 Results 109
    9.6.1 Ecological preferences 109
    9.6.2 Environmental analysis 113
    9.6.3 Mass extinction analysis 115
9.7 Discussion and conclusions 121
9.8 Acknowledgements 124
References 125

10 K-Pg mass extinction 129
Ashraf M. T. Elewa
References 130

11 Causes of mass extinction at the K/Pg boundary: A case study from the North African Plate 133

Ashraf M. T. Elewa and Ahmed M. Dakrory
11.1 Introduction 133
11.2 Material and methods 134
11.3 Quantitative results 135
    11.3.1 Cluster analysis 135
    11.3.2 Principal coordinate analysis 137
11.4 Qualitative results 138
11.5 Discussion and conclusions 142
11.6 Acknowledgements 146
References 147

12 Patterns and causes of mass extinction at the K/Pg boundary: Planktonic foraminifera from the North African Plate ... 149

Ashraf M. T. Elewa and Ahmed M. Dakrory
12.1 Introduction 149
12.2 Stratigraphy 150
    12.2.1 Sudr Chalk 150
    12.2.2 Dakhla Formation 151
12.3 Material and methods 151
12.4 Results and discussion 153
12.5 Acknowledgements 157
References 158

13 Quaternary extinctions in Southeast Asia 159

Julien Louys
13.1 Introduction 159
13.2 The Quaternary "megafauna" extinctions 160
    13.2.1 What are megafauna? 160
    13.2.2 The debate 160
    13.2.3 Towards a reconciliation 164
13.3 Quaternary Extinctions in Southeast Asia 165
    13.3.1 Geography of Southeast Asia 165
    13.3.2 Geological History 166
    13.3.3 Climate 168
    13.3.4 Vegetation 168
    13.3.5 Sea level changes 170
13.4 Southeast Asia's megafauna 172
    13.4.1 Dubois's Antelope 172
    13.4.2 Yunnan Horse 172

    13.4.3 Asian Gazelle 173
    13.4.4 Giant Ape 173
    13.4.5 Robust Macaque 173
    13.4.6 Giant Hyena 174
    13.4.7 Rhinoceroses 174
    13.4.8 Pigs 174
    13.4.9 Stegodons 175
    13.4.10 Malayan Tapir 175
    13.4.11 Giant Tapir 176
    13.4.12 Serow 176
    13.4.13 Giant Panda 176
    13.4.14 Asian Spotted Hyena 176
    13.4.15 Archaic elephant 177
    13.4.16 Orangutan 177
  13.5 Human overhunting in Southeast Asia? 177
  13.6 Climate change and megafauna 178
  13.7 The modern extinction crisis 181
  13.8 Summary 182
  13.9 Acknowledgements 183
  References 183

14 Current mass extinction 191
  Ashraf M. T. Elewa

15 Current insect extinctions 195
  Panos V. Petrakis
  15.1 Introduction 195
  15.2 Insect mass extinctions in the past 197
  15.3 Types of current insect extinctions 200
    15.3.1 Lessons from island ecology 202
    15.3.2 Insect extinctions induced by changes in
           succession status 204
    15.3.3 Insect extinctions from climate change 206
  15.4 The prediction of certain insect extinctions 209
    15.4.1 The Darwin-Lyell extinction model 209
    15.4.2 The Raup extinction model and the kill curve 210
    15.4.3 Causes of current insect extinction 212
    15.4.4 Climate change 212
    15.4.5 Habitat fragmentation, destruction, modification 217
    15.4.6 Species invasions/introductions 221
    15.4.7 Co-extinctions 223
    15.4.8 Hybridization and introgression 225

15.5 Conservation of insect species and their biodiversities 227
   15.5.1 Estimating conservation priority 227
   15.5.2 Conserving insects through habitat/plant
          conservation 234
   15.5.3 Conserving insect biodiversity in city parks and
          road verges 237
15.6 Acknowledgements 240
References 240

Index **251**

1 Mass Extinction - a general view

Ashraf M. T. Elewa

Geology Department, Faculty of Science, Minia University, Minia 61519, Egypt, aelewa@link.net

Mass extinction is considered as the most subject matter in paleontology that received several debated input. When looking to the past, we find five major mass extinctions, I believe, in the fossil record (e.g. Late Ordovician, Late Devonian, Late Permian, Late Triassic and Late Cretaceous). Some authors believe in six mass extinctions by adding Cambrian to the previous five events, some others speak on cycles of mass extinctions up to 23 events since the Cambrian. Nevertheless, patterns and causes of these mass extinctions still disputable. It is notable, however, that there is a variety in degree of diversity loss between the minor and major biotic disaster. Yet, the scientists stress on the five major mass extinctions with more focus on the Cretaceous/Paleogene event. This focus is normally due to the extinction of dinosaurs during this interval, which can make the cover of Time magazine, but not productids or fusulinids as stated by Prothero (1998).

The ordinary questions of the scientists interested in this field are lying in the search for common causes and patterns that may lead to a general theory of extinction. The most frequent arguments are related either to bad genes or to bad luck. Whatever were the causes, it is logic to give an idea on each of the five major mass extinctions in the fossil record.

The first major mass extinction event started in the Late Ordovician time. Despite the rigorous crisis affected this extinction, for no reason it has so far received little attention from scientists. Sepkoski (1989) affirmed that 57% of the marine genera disappeared in this crisis (some authors stated that it is the second biggest extinction of marine life, ranking only below the Late Permian extinction). Scientists assigned this extinction to global cooling that triggered glaciation and significant lowering of the sea level. As a result, one hundred families of marine invertebrates died, including two-thirds of all brachiopod and bryozoan families, and once-flourishing trilobites as well as archaic groups of echinoderms also died out (for more information see Sepkoski 1984, 1989; Hallam and Wignall

1997; Prothero 1998). It is notable that scientists excluded the single extraterrestrial iridium anomaly from the causes of this big mass extinction.

The Late Devonian extinction event was almost as remarkable, obliterating 75% of the species and 50% of the genera in the marine realm (McGhee 1995). Some specialists clued that 19% of all families have been died and went extinct during this event. Again global cooling was the most important factor affected this mass extinction according to different authors. On the other hand, the effect of extraterrestrial impact on this event is strongly debated.

The third major mass extinction occurred in the Late Permian-Early Triassic and is considered as the biggest within the five major extinction events. In this event, about 57% of all families, 83% to 90% of all genera, and 96% of the species in the marine realm have wiped out. Even though, the causes of this biggest mass extinction still controversial. Benton (2003) assigned this event to the rise of carbon in the atmosphere. Some others referred to global warming and a depletion of oxygen in the atmosphere due to massive volcanic eruptions in Siberia. A third group blames an asteroid impact, and the fourth group suggested global cooling. Erwin (1993) and Prothero (1998) believe that a combination of different disasters caused this extinction.

The Late Triassic extinction event is considered as the smallest of the big five extinction events. In this event about 23% of all families and 48% of all genera went extinct (Sepkoski 1989). Most authors assigned this event either to global cooling or to volcanic activity.

The fifth major mass extinction started in the Late Cretaceous. It eliminated about 17% of all families and more than 50% of all genera. No doubt, it is the most interesting and attractive event in the fossil record, as it wiped out the dinosaurs. As Prothero (1998) said, prior to 1980 there were a number of ideas for the extinction of the dinosaurs (cooling, warming, disease, inability to digest angiosperms, mammals ate their eggs). I add to this the possibility of dinosaurs to mistakenly kill themselves by their rear spines. However, all these causes, of course, did not kill other groups of organisms!! Therefore, scientists searched for other reasons to this mass extinction, which is apparently started in 1980 when Alvarez et al. suggested that this extinction is due to an extrateresstrial bolide impact with about 10 km in diameter. Since then, controversial evidences have been proposed for this extinction with two significant schools. The first one presumes the effect of the bolide impact (catastrophe), whereas the second believes in gradual extinction (volcanic activity, environmental and climatic changes). This situation changed in 1996, when Molina et al. supposed that both catastrophic and gradual extinctions might have occurred. Elewa and Dakrory (submitted) ensured the results of Molina et al. (1996).

After all, it seems that the sixth major mass extinction is underway. If this is true what are the causes and evidences? The book in hand may give important information on this area under discussion.

In summary, usually, most of the published books on the subject just focus on restricted groups of organisms and could not answer several questions relating to mass extinction. However, the present book is different in combining two main aspects:

1. Five major mass extinctions of the fossil record; and more importantly
2. Contributions on minor extinctions and current mass extinction

These two aspects are introduced through interesting studies of mass extinctions in diverse organisms ranging from small invertebrates to big vertebrates, and take account of the most admired subjects discussing mass extinctions in attractive groups of organisms like insects and dinosaurs.

Moreover, as in my previously edited books with Springer, I selected an exceptional group of specialists working on this phenomenon to explain and write about the subject.

Regarding addressees, mass extinction is one of the most popular topics for students, at all levels, researchers, and professionals. Besides, this book project represents advanced ideas and useful synopsis on this subject stuff.

Finally, I would like to thank Dr. P. David Polly (USA) for writing the foreword. The contributors and the publishers of Springer-Verlag are deeply appreciated. As usual, I am much indebted to the staff members of Minia University of Egypt.

References

Alvarez LW, Alvarez W, Asaro F, Michel HV (1980) Extraterrestrial cause for the Cretaceous-Tertiary extinction. Science 208: 1095-1108

Benton MJ (2003) When life nearly died: The greatest mass extinction of all time. Thames and Hudson

Erwin DH (1993) The great Paleozoic crisis: Life and death in the Permian. Critical Moments in Paleobiology and earth History Series, Columbia University Press, New York

Hallam A, Wignall PB (1997) Mass extinctions and their aftermath. Oxford University Press, Oxford

McGhee Jr GR (1995) The Late Devonian mass extinction: The Frasnian/Famennian crisis. Columbia University Press, New York

Molina E, Arenillas I, Arz JA (1996) The Cretaceous/Tertiary boundary mass extinction in planktic foraminifera at Agost, Spain. Rev Micropaléont 39 (3): 225-243

Prothero DR (1998) Bringing fossils to life: An introduction to paleobiology. WCB/McGrow-Hill, USA, 560 pp

Sepkoski Jr JJ (1984) A kinetic model of Phanerozoic taxonomic diversity, III. Post-Paleozoic families and mass extinctions. Paleobiology 10 (1984): 246-267

Sepkoski Jr JJ (1989) Periodicity in extinction and the problem of catastrophism in the history of life. J Geol Soc London 146 (1989): 7-19

2 Late Ordovician mass extinction

Ashraf M. T. Elewa

Geology Department, Faculty of Science, Minia University, Minia 61519, Egypt, aelewa@link.net

The Ordovician period was an era of extensive diversification and expansion of numerous marine clades. Although organisms also present in the Cambrian were numerous in the Ordovician, a variety of new types including cephalopods, corals (including rugose and tabulate forms), bryozoans, crinoids, graptolites, gastropods, and bivalves flourished. Ordovican communities typically displayed a higher ecological complexity than Cambrian communities due to the greater diversity of organisms. However, as in the Cambrian, life in the Ordovician continued to be restricted to the seas. The Ordovician extinction occurred at the end of the Ordovician period, about 440-450 million years ago. This extinction, cited as the second most devastating extinction to marine communities in earth history, caused the disappearance of one third of all brachiopod and bryozoan families, as well as numerous groups of conodonts, trilobites, and graptolites. Much of the reef-building fauna was also decimated. In total, more than one hundred families of marine invertebrates perished in this extinction. The Ordovician mass extinction has been theorized by paleontologists to be the result of a single event; the glaciation of the continent Gondwana at the end of the period. Glacial deposits discovered by geologists in the Saharan Desert provide evidence for this glaciation event. This glaciation event also caused a lowering of sea level worldwide as large amounts of water became tied up in ice sheets. A combination of this lowering of sea-level, reducing ecospace on continental shelves, in conjunction with the cooling caused by the glaciation itself are likely driving agents for the Ordovician mass extinction (Dr. Ken Hooper Virtual Paleontology Museum, Ottawa-Carleton Geoscience Center and Department of Earth Sciences, Carleton University, Ontario, Canada).

Another information can be obtained from Wikipedia, the free encyclopedia, which considered the Ordovician-Silurian extinction event as the second largest of the five major extinction events in Earth's history in terms of percentage of genera that went extinct. The Wikipedia assigned this event to either glaciation, as it is believed by many scientists, or to the gamma ray burst originating from an exploding star within 6,000 light years of Earth (within a nearby arm of the Milky Way Galaxy). A ten-second burst would have stripped the Earth's atmosphere of half of its ozone almost immediately, causing surface-dwelling organisms, including those responsible for planetary photosynthesis, to be exposed to high levels of ultraviolet radiation. This would have killed many species and caused a drop in temperatures. While plausible, there is no unambiguous evidence that such a nearby gamma ray burst has ever actually occurred.

The Peripatus website clued on the net that mass extinctions of tropical marine faunas occurred at the end of the Ordovician when 100 or more families became extinct, including more than half of the bryozoan and brachiopod species. This website cited the following possible causes for this event:

- climatic cooling
- major glaciation
- sea level drop
- Iapetus Ocean (proto-Atlantic) closed, eliminating habitats
- Cambro-Ord platform collapsed
- Taconic Orogeny

On the other hand, the University of Southern California, Department of Earth Sciences, considered the mass extinction of organisms at the end of the Ordovician as propably the greatest mass extinction ever recorded in the Earth history with over 100 families going extinct. The Department of Earth Sciences noted the following:

1. One idea was that it was the breakup and movement of the large super continent into many fragments. However, modern biology teaches us that this would not likely lead to extinctions, rather it would provide additional niche space for groups to expand into.
2. The more likely cause is that the Earth cooled, particularly the oceans where most of the organisms lived during the Ordivician (Remember there were not land plants and no evidence of land organisms yet). All the extinctions occurred in the oceans.

Paul Recer, in the Associated Press (2004), confirmed that the Late Ordovician mass extinction event was the second-largest extinction in the Earth's history, the killing of two-thirds of all species, may have been caused by ultraviolet radiation from the Sun after gamma rays destroyed the Earth's ozone layer. This supports the gamma ray burst hypothesis.

Prothero (1998) avowed that warm adapted taxa seem to be the chief victims, suggesting that a global cooling event might have been responsible. Wilde and Berry (1984) introduced the possibility of that the cooling and glaciation regression may brought the biologically toxic waters to the surface leading to severe conditions on the sensitive shallow marine benthic community. Orth (1989) declared that no clear evidence of a single extraterrestrial iridium anomaly has yet been documented for the Ordovician.

This short essay clarifies that the causes of the Late Ordovician mass extinction event are still debated, and no scientific team could give stronger evidences than the other team.

References

Wilde P, Berry W (1984) Destabilization of the oceanic density structure and its significance to marine "extinction" events. Palaeogeogr, Palaeoclimat, Palaeoecol 48: 143-162

Orth CJ (1989) Geochemistry of the bio-event horizons. In Donovan SK (ed) Mass Extinctions – Processes and Evidence. New York, Columbia University Press, 266 pp

Prothero DR (1998) Bringing fossils to life: An introduction to paleobiology. WCB/McGrow-Hill, USA, 560 pp

3 The End Ordovician; an ice age in the middle of a greenhouse

Curtis R. Congreve

Department of Geology, University of Kansas, Lawrence, KS 66045, USA

3.1 Introduction

With millions of years of Earth history to study, it is interesting that so much attention is devoted to the rare and relatively short lived time intervals that represent Earth's major mass extinctions. Perhaps this interest is twofold. On the one hand, there is a fair degree of self-interest in studying extinction considering the present biodiversity crisis we now face. On the other hand, these periods of time have had an incredible effect on life history. These cataclysmic times represent periods of environmental and ecological abnormality amidst millions of years of relative stability. As such, these mass extinctions are times of incredible change, which can be studied both evolutionarily as well as ecologically. When viewed through an evolutionary framework, mass extinction events represent unique time periods in the history of life. These ecological crises prune the tree of life, removing families and killing off entire lineages at random (Raup 1981). Those lineages lucky enough to survive the catastrophe continue and diversify. Often, it is by this seemingly random removal of organisms that large scale evolutionary changes can take place. Take, for instance, the present state of our world, which is primarily dominated by large mammals. Had the non-avian dinosaurs not met with an untimely demise at the end of the Cretaceous, mammals would probably never have been able to diversify into the numerous forms that we see today. It is for this reason that the study of mass extinction events is incredibly important to evolutionary biology. Mass extinctions are essentially historical "turning points" that affect the evolution of all of the Earth's biota on a gross scale.

Mass extinctions can also be studied as ecological experiments. Ultimately, mass extinctions represent times of ecological upheaval in which climate may shift and ecological niche space can be destroyed. By studying both the causes of these ecological perturbations, as well as the affect that these changes have on the biota, we are able to better understand how life reacts under times of ecological stress. This in turn can help us predict the patterns that we might expect in future mass extinctions. This type of study is of particular importance in our present biodiversity crisis.

The end Ordovician mass extinction is a unique time period that offers a great deal of study material to geologists interested in both the ecological and evolutionary aspects of mass extinctions. The end Ordovician mass extinction is a time of great ecological upheaval. The cause of this massive die off has long been considered to be a glacial period (Berry and Boucot 1973; Sheehan 1973). Although this interpretation appears to be quite sound, there is still a great deal of debate about the timing of the glacial event as well as its forcing mechanism. The original interpretation proposed by Berry and Boucot (1973) was that the glacial period might have lasted millions of years and that global cooling was gradual. Recent evidence (Melott et al 2005; Brenchley et al. 1994) suggests that the glaciation was incredibly sudden and brief, possibly lasting only a few hundred thousand years. Furthermore, it appears that this glacial period occurred in the middle of a greenhouse climate. The extinction patterns in the end Ordovician glacial period are also intriguing, especially the patterns found in trilobites. Trilobite species with cosmopolitan biogeographic ranges preferentially go extinct while more endemic species are more prone to survive (Chatterton and Speyer 1989). This is contrary to the pattern frequently identified by Stanely (1979), Vrba (1980), Eldridge (1979) and others, who argued that organisms with larger biogeographic ranges tend to have lower extinction rates than those with smaller, more endemic ranges. Yet, in the Ordovician extinction it is the endemic species that tend to survive.

This paper will focus on previous research that has been conducted on the Ordovician mass extinction. Furthermore, several of the major unresolved issues concerning the causes of the glaciation as well as the patterns of the extinction will be emphasized; this paper will conclude with a discussion of new research that hints at a possible forcing mechanism for the sudden onset of glaciation.

3.2 Early research: The discovery of the glacial period

Some of the first scientists to invoke a massive glacial period at the end of the Ordovician were Berry and Boucot (1973). Berry and Boucot were interested in explaining a global pattern within the sedimentary record. During the early Silurian there was substantial evidence of onlap deposits. Prior to this rapid rise of sea level, there is some evidence (Kielan 1959) that the sea level had been steadily dropping during the late Ordovician.

What could have resulted in this global fall and rise of sea level? One explanation could have been tectonic processes, such as orogenic events. These processes could raise and lower the land, thus changing the land's position relative to the sea level. However, in order for this mechanism to result in a seemingly global sea level rise, there would need to be synchronicity amongst all tectonic events occurring on the planet. Berry and Boucot (1973) did not find any significant time correlation across regions between the tectonic events that occurred during the end Ordovician. Thus, another mechanism needed to be invoked in order to explain this global phenomenon.

Again, the clues to discovering this mechanism came from studying the sedimentological record. During the late Ordovician, gravel and cobble deposits were found in North Africa, which were interpreted as being glacially derived sediments (Beuf et al 1971; Destombes 1968; Dow et al. 1971). Furthermore, late Ordovician age sedimentary deposits were found in Europe that were interpreted as being ice rafted debris (Arbey and Tamain 1971; Dangeard and Dore 1971; Shönlaub 1971). These sedimentary deposits suggested that there might have been an increase in glacial ice during the late Ordovician. Since the presence of this ice correlated with the estimated time of sea level fall, Berry and Boucot (1973) proposed that massive glaciation was the mechanism responsible for the drop in sea level. The concept behind this theory is similar to a phenomenon which occurred during the recent Pleistocene glaciations: Newell and Bloom (1970) (to cite a recent reference) observed that during the last glacial period the sea level was approximately 100 meters lower than it is at present. This is because ice that rests on land effectively traps water and prevents it from reaching the ocean. As more land locked ice builds up, it traps more water from reaching the oceans and the sea level falls. This is the mechanism that Berry and Boucot (1973) invoked to explain the sedimentological pattern observed at the end Ordovician. During the late Ordovician, the onset of a glacial period resulted in the lowering of global sea level as water was trapped in continental glaciers. As the glaciers melted, the water was returned to the oceans and sea level rose. This explained the onlap deposits found in the early Silurian. Berry and Boucot (1973) concluded that this process was probably very gradual, and that the end Ordovician glacial period lasted millions of years, unlike the recent Pleistocene glaciations.

This glacial process was supported by Sheehan (1973) who cited a biogeographic pattern of brachiopod evolution that he deemed consistent with glacially driven eustatic changes. Prior to the end Ordovician, there existed two major brachiopod provinces, a North American province and an Old World Province. After the extinction event, the North American province was gone and was replaced by species that were derived from the Old World faunas. Sheehan (1973) believed that this faunal interchange, as well as the extinction of the North American fauna, was caused by eustatic sea level changes during the glacial event. Prior to the glaciation, shallow epicontinental seas (approximately 70 meters deep) covered much of North America (Foerste 1924). These epicontinental seaways represented the habitat for the North American brachiopod fauna. During the 100 meter sea level drop proposed by Berry and Boucot (1973), these epicontinental seaways would have almost entirely dried up. Such a massive reduction in habitat space would have greatly stressed the North American brachiopods, ultimately resulting in their extinction. This habitat space would then have been repopulated by the nearby Old World fauna, which would have been less affected by the extinction because the higher European topography meant that the Old World brachiopods were adapted to shelf niche space and not epicontinental seaways (Sheehan 1975). Sheehan envisioned this process as being gradual, with the North American faunas going extinct over the course of the glacial period and the Old World faunas steadily replacing and out competing the local fauna (1973, 1975). However, he admitted that biostratigraphy of the Late Ordovician period was poor and thus any time correlation must be taken with a grain of salt.

3.3 Fast or slow: The changing face of the Ordovician glaciation

During the next twenty years, there was a great deal of research concerning the timing of the glacial onset as well as how long the glacial period lasted. Originally, the glaciation was thought to have started in the Caradoc and continued into the Silurian. However, this estimated glacial duration met with a fair degree of contention. The Caradoc had originally been established as the onset of glaciation because of faunal assemblages found in glacial sequences in the Sahara (Hambrey 1985). However, these assemblages had been described as being older preglacial clasts that had been ripped from the bedrock and incorporated into the glacial sediments (Spjeldnases 1981), thus they could not be used to date the sequence. Crowell (1978) had suggested that the glacial period extended far into the Silurian. This conclusion was based on tillite deposits found in South America that were believed to be Wenlock in age (Crowell 1978). However, Boucot (1988) called this age constraint into question, citing that the paleontological record in the area was insufficient for use in biochronology. Furthermore, he suggested that the tillites were probably from the Ashgill.

An Ashigillian date for the glacial episode was further corroborated by two other pieces of evidence. First, Brenchley et al (1991) identified Ashgillian age glacial-marine diamictites that were interbedded with fossiliferous deposits. Second, Brenchley et al (1994) conducted a global geochemical study analyzing $\delta^{18}O$ and $\delta^{13}C$ of brachiopod shells found in the midwestern United States, Canada, Sweden, and the Baltic states. They were unable to consistently use brachiopods of the same genera and instead used a wide variety of species but found that their data clustered together relatively well. This helped to ensure that any pattern they found in their data was an actual signal and not just error caused by varying biotic isotopic fractionation. The results of this study showed that there was a sharp positive increase in $\delta^{18}O$ during the Ashgill. $\delta^{18}O$ concentrations returned to their pre-Ashgillian state at the end of the Ordovician. This increase in $\delta^{18}O$ concentration is consistent with what would be expected from a glacial event. Global cooling and accumulation of negative $\delta^{18}O$ ice would cause global ocean water to become enriched in ^{18}O, resulting in the positive shift in $\delta^{18}O$. Once the ice melted and the temperatures returned to normal, the $\delta^{18}O$ concentration returned back to its pre-Ashgillian state. This geochemical evidence indicates that the onset of glaciation occurred during the Ashgill and that the glacial period was incredibly brief. But what could have caused this glaciation?

The results of Brenchley et al (1994) become even more peculiar when you take into account paleoclimatic studies of the Ordovician and Silurian. Research indicates that the atmospheres of the late Ordovician and early Silurian had very high concentrations of CO_2 (Berner 1990, 1992; Crowley and Baum 1991). High concentrations of CO_2 would act to keep the climate of the late Ordovician in a greenhouse condition. How could a glacial period exist in the middle of a greenhouse? Brenchley et al (1994) proposed one possible mechanism that was consistent with their $\delta^{13}C$ data. When the $\delta^{18}O$ data shifts towards the positive, there is a contemporaneous shift in the $\delta^{13}C$ towards the positive as well. This shift in $\delta^{13}C$ was envisioned as an increase in marine productivity because of increased cool deepwater production. Before the onset of glaciation, the deepwaters of the Ordovician would have been warm and poorly circulated (Railsback et al. 1990). If global temperatures cooled, the ocean water would have cooled as well which would help to increase oceanic circulation. This would have made the oceans rich in nutrients and increased the productivity of the oceans, which in turn would act to remove CO_2 from the atmosphere, effectively lowering the Earth's temperature and allowing for the brief icehouse conditions to occur (Brenchley et al. 1994). Although this theory helps to explain why a glacial period could persist in the midst of greenhouse conditions, it still requires that some initial forcing mechanism act to cool the Earth's temperature. The forcing mechanism that was cited by Brenchley et al. (1994) was the migration of Gondwana. As the continent migrated pole-ward, it would have accumulated ice and snow, thus increasing the Earth's albedo and decreasing global temperature (Crowley and Baum 1991). However, there is a problem associated with this mechanism. The migration of Gondwana is a tectonic forcing mechanism, and tectonism usually operates on million year time scales. Even if the onset of glaciation was somehow sudden (if the Earth needed a threshold albedo value to spontaneously glaciate), it would still take millions of years until the glacial period ended. This does not coincide with the brief glacial period proposed by Brenchley et al (1994). Thus, it seems counterintuitive for the migration of Gondwana to be the initial forcing mechanism for the glaciation.

A recent study by Melott et al (2004) proposes that the Ordovician glaciation could have been caused by a gamma ray burst (GRB). Such an event could result in a sudden and brief glacial period on the order of time that is predicted by Brenchley et al (1994). The theory is as follows: A GRB from a nearby star sends high-energy waves in the form of photons out into space. These high-energy waves make it to Earth and begin initiating various atmospheric reactions. The net effect of these reactions is twofold. First, the increased cosmic radiation would destroy ozone, thus thinning the planet's ozone layer. Second, there would be increased production of NO_x gases. These opaque gases would build up in the atmosphere, darkening the Earth's skies and preventing sunlight from reaching its surface. This build up of NO_x gases would result in global cooling. Melott et al (2004) estimated that the GRB would have lasted only a matter of seconds, but the effects that it would have had on the atmosphere would have taken years to equilibrate (Laird et al 1997). This theory is very interesting because it offers a mechanism by which the Ordovician glaciation could have occurred suddenly during greenhouse conditions. Furthermore, it explains why the glacial period was so brief. After the GRB event was over, the NO_x gases in the atmosphere responsible for global cooling began to slowly decay over the course of several years. However, the effects of this initial cooling caused by the GRB probably contributed to other factors which helped to prolong global cooling, such as increased albedo due to ice accumulation, or the increased ocean productivity due to increased circulation as proposed by Brenchley et al (1994). This ultimately would have resulted in the brief and unstable icehouse conditions at the end Ordovician. The GRB hypothesis might also explain some of the extinction patterns during the Ordovician extinction, in particular those pertaining to trilobites.

3.4 Trilobite extinction and larval form

Chatterton and Speyer (1989) drew attention to an unexpected pattern associated with the late Ordovician extinction. They studied trilobite extinction patterns and related survivability to the proposed lifestyle and larval forms of each family. What they discovered was that the greater the duration of an inferred planktonic larval phase, the greater the probability of extinction. Trilobites that were inferred to have planktonic larval stages and benthic adult stages were more likely to go extinct than trilobites that spent their entire lives in a benthic stage. Furthermore, trilobites that were worst hit by the extinction (and subsequently entirely wiped out) were those organisms that had an inferred pelagic adult stage. This pattern may be the opposite of what we might tend to expect: Species with planktonic larval stages or pelagic adult stages would tend to have larger biogeographic ranges than species that are purely benthic. As such, these planktonic or pelagic trilobites would have tended towards being more ecologically generalized, whereas the benthic species would have tended to be more specialized and endemic. Organisms that are ecological generalists and have broad geographic ranges usually have very low extinction rates, whereas specialists tend to have very high extinction rates (Vrba 1980). Therefore, it would be natural to assume that generalists would be better buffered against extinction than specialists. However, in the end Ordovician it is the more narrowly distributed putative specialist organisms that are best suited to survival, while the more broadly distributed putative generalists are more at risk. Chatterton and Speyer (1989) explained this pattern as being the result of a trophic cascade resulting from the effects of global cooling. As the ocean temperatures cooled during the glacial event, the lower water temperatures would have eventually acted to reduce the productivity of phytoplankton (Kitchell 1986, Kitchell et al. 1986, Sheehan and Hansen 1986). Since the plankton was the basis for the food chain, there would have been increased extinction up the trophic levels in planktonic and pelagic organisms. Benthic trilobites would have been buffered from the effects of this trophic cascade scenario because they were probably detritus feeders who would have eaten the remains of the dead pelagic and planktonic organisms.

Although this is one possible scenario that could have resulted in this extinction pattern, another explanation emerges if we view the extinction as being caused by a GRB as proposed by Melott et al. (2004). One of the proposed effects of a GRB is thinning of the ozone layer. If the ozone layer thinned during the end Ordovician, this would have allowed a larger flux of high-energy ultraviolet (UV) radiation to reach the surface of the planet, increasing rates of deadly mutations. Organisms that lived at the surface of the oceans or high up within the water column would have been more affected by this increase in UV radiation than benthic organisms that would have been better shielded by surrounding sediments. Therefore, the planktonic larval forms and pelagic trilobites would have already been under much more stress than their benthic counterparts at the onset of glaciation, possibly even before major global cooling had set in. The increase of high-energy UV radiation reaching the Earth's surface, coupled with the sudden glacio-eustatic changes and global cooling would have hit the Earth's biota in a devastating one-two punch.

I propose a third process that might also help to explain the trilobite extinction pattern observed at the end Ordovician. Vrba (1993, 1995) has shown that fluctuations in paleoclimate could result in speciation. According to Vrba (1993, 1995) as climates change, the organisms that live within their respective climatic ranges will track their preferred climate. In times of extreme climate change, such as the onset of an icehouse condition, the species ranges of tropical and temperate species would begin to shrink and move towards the equator. As the species ranges shrink, there is a greater probability that small populations could become reproductively isolated from the main population. If this situation persists for long enough, these small populations will speciate by means of allopatric speciation. Thus, somewhat paradoxically, the habitat destruction caused by massive global climate change could also act to temporarily increase levels of speciation. Applying this theory to the end Ordovician, we would expect that as global cooling shrunk the biogeographic ranges of trilobites, they too would experience an increase in speciation rate that might have helped them to stave off the heightened extinction rates. Endemic species such as the trilobites with benthic larval stages perhaps would have been easier for smaller populations to become reproductively isolated by habitat destruction due to the specificity of their environmental constraints. On the other hand, generalist species might have been more difficult to reproductively isolate long enough to result in speciation. The net result would be that generalist trilobites would have been given less of a boost to their speciation rate during the glacial episode than endemic species and would therefore have been less buffered against the effects of the raised extinction rates. A detailed study of extinction and speciation rates of planktonic larval and non-planktonic larval trilobites over the course of the Ordovician would be necessary in order to test this hypothesis.

3.5 Future work

The Ordovician mass extinction event is a unique time period that offers a great deal of research opportunities in all fields of geology. The question of the forcing mechanism for the end Ordovician glacial period is still a matter of debate, but Melott et al. (2004) have provided an interesting new way of viewing the extinction event. Future research could focus on testing patterns of extinction in greater detail within the context of a GRB scenario to see if these patterns are indeed congruent with the sudden onset glaciation and increased high-energy radiation expected during a GRB. One method of testing this is by studying biogeographic patterns using Geographic Information Systems (GIS) (ESRI 2005) throughout the Ordovician. Since organisms will track their preferred environment through time, climatic changes should leave an imprint on the biogeographic record that could be analyzed. I am currently conducting a project using GIS to study Cyclopygid trilobite distribution throughout the extinction. The time of initial global cooling, as well as the rate at which the glaciation advanced, could be determined by studying biogeographic patterns. If the ice age was brought on over the course of a million or so years by gradual processes such as the migration of Gondwana, I would expect to see the gradual shrinking of tropical species ranges closer to the equator as global temperatures cooled. If the glaciation was much more sudden, like the proposed GRB scenario, the shrinking of species ranges would probably not be preserved within the fossil record. Instead, the pattern that might be expected would be a sudden change, akin to flipping a switch. I propose that a detailed biogeographic analysis that incorporates modern techniques such as a phylogenetic biogeographic study and GIS could help to illuminate the possible causes of the Ordovician glacial period.

3.6 Acknowledgements

I would like to thank Bruce S. Lieberman (BSL) for comments on an earlier version of this manuscript, and also NASA Astrobiology (NNGOYGMYIG to BSL) for support of this research.

References

Arbey, Tamain (1971) Existence d'une glaciation siluro-ordovicienne en Sierra Morena (Espagne): Acad Sci Comptes 272: 1721-1723

Berner RA (1990) Atmopsheric carbon dioxide levels over Phanerozoic time. Science 249: 1382-1386

Berner RA (1992) Palaeo-CO_2 and climate, Nature 358: 114

Berry WBN, Boucot AJ (1973) Glacio-Eustatic Control of Late Ordovician-Early Silurian Platform Sedimentation and Faunal Changes. Geological Society of America Bulletin 84 (1): 275-284

Beuf S, Biju-Duval B, DeCharpal O, Rognon P, Garriel O, Bennacef A (1971) Les gres du Paleozoique inferieur au Sahara-sedimentation et discontinuities evolution structurale d'un craton. Inst Francais Pétrole Rev 18: 464

Boucot AJ (1988) The Ordovician-Silurian boundary in South America. British Museum of Natural History (Geology) Bulletin 43: 285-290

Brenchley PJ, Romandno M, Young TP, Storch P (1991) Hirnantian glaciomarine diamictites- Evidence for the spread of glaciation and its effect on Ordovician faunas. In Barnes CR, Williams SH (eds) Advances in Ordovician geology. Geological Survey of Canada Paper 90-9: 325-336

Brenchley PJ, Marshall JD, Carden GAF, Robertson DBR, Long DGF, Meidla T, Hints L, Anderson TF (1994) Bathymetric and isotopic evidence for a short-lived Late Ordovician glaciation in a greenhouse period. Geology 22: 295-298

Chatterton BDE, Speyer SE (1989) Larval ecology, life history strategies, and patterns of extinction and survivorship among Ordovician trilobites. Paeobiology 15 (2): 118-132

Crowell JC (1978) Gondwana glaciations, cyclothems, continental positioning and climatic change. American Journal of Science 278: 1345-1372

Crowley TJ, Baum SK (1991) Towards reconciliation of Late Ordovician (~440 Ma) glaciation with very high CO_2 levels. Journal of Geophysical Research 98: 22, 597-22, 610

Dangeard L, Dore F (1971) Facies glaciares de l'Ordovicien Superieur en Normandie. Bur. Recherches Géol. Et Minières Mem 73: 119-127

Destombes J (1968) Sur la nature glaciare des sediments du groupe du 2e Bani: Ashgill superieur de l'Anti Atlas. Acad Sci Comptes Rendus 267: 684-689

Dow DB, Beyth M, Hailu T (1971) Paleozoic glacial rocks recently discovered in northern Ethiopia. Geol Mag 108: 53-59

Eldredge N (1979) Alternative approaches to evolutionary theory. Bull Carn Mus. Nat Hist 13: 7-19

Foerste AF (1924) Upper Ordovician faunas of Ontario and Quebec. Candian Geol Surv Mem 138: 1-255

Hambrey MJ (1985) The Late Ordovician-Silurian glacial period. Palaeogoegraphy, Palaeoclimatology, Palaeoecology 99: 9-15

Kielan Z (1959) Upper Ordovician trilobites from Poland and some related forms from Bohemia and Scandinavia. Pal Polonica 11: 198

Kitchell JA (1986) The selectivity of mass extinction: causal dependency between life-history and survivorship. Fourth North American Paleontological Convention, Abstracts: A25

Kitchell JA, Clark DL, Gombos AMJr (1986) Biological selectivity of extinction: a link between background and mass extinction. Palaios 1: 504-511

Laird CM, Gehrels N, Jackman CH, Chen W (1997) Modeling supervovae effects on the Earth's atmosphere: implications for ozone and odd nitrogen levels. Trans Amer Geophys Union 79 (17): April 29, S91

Melott, A. L., Lieberman, B. S., Laird, C. M., Martin, L. D., Medvedev MV, Thomas BC, Cannizzo JK, Gehrels N, Jackman CH (2004) Did a gamma-ray burst initiate the late Ordovician mass extinction? International Journal of Astrobiology 3 (1): 55-61

Newell ND, Bloom AL (1970) the reef flate and "Two-meter eustatic terrace" of some Pacific atolls. Geological Society of America Bulletin 81: 1881-1894

Railsback LB, Ackerley SC, Anderson TF, Cisne JL (1990) Paleontological and isotope evidence for warm saline deep waters in Ordovician oceans. Nature 343: 156-159

Raup DM (1981) Extinction: bad genes or bad luck? Acta Geologica Hispanica 16: 1-2: 25-33

Sheehan PM (1973) The relation of Late Ordovician glaciation to the Ordovician-Siluarian changeover in North American brachiopod fauna. Lethia 6 (2): 147-155

Sheehan PM (1975) Brachiopod synecology in a time of crisis (Late Ordovician-Early Silurian). Paleobiology 1: 205-212

Sheehan PM, Hansen TA (1986) Detritus feeding as a buffer to extinction at the end of the Cretaceous. Geology 14: 868-870

Shönlaub HP (1971) Palaeo-environmental studies at the Ordovician/Silurian boundary in the Carnic Alps. Bur Recherches Géol Et Minières Mem 73: 367-378

Spjeldnases M (1981) Lower Palaeozoic climatology. In Holland CH (ed) Lower Palaeozoic of the Middle East, eastern and southern Africa, and Antarctica. Chichester, United Kingdom, Wiley 199-256

Stanley SM (1979) Macroevolution; pattern and process. San Francisco: WH Freeman

Vrba ES (1980) Evolution, species and fossils: how does life evolve? S Afr J Sci 76: 61-84

Vrba ES (1993) Turnover-pulses, the Red Queen, and related topics, Amer J Sci 293: 418-452

Vrba ES (1995) Species as habitat-specific, complex systems. In Lambert DM, Spencer HG (eds) Speciation and the Recognition Concept. Johns Hopkins University Press, Baltimore, 3-44

4 Silurian global events – at the tipping point of climate change

Mikael Calner

GeoBiosphere Science Centre, Lund University, Lund, Sweden, mikael.calner@geol.lu.se

4.1 Introduction

Mass extinction events affect a wide breadth of ecosystems and are one of the major driving mechanisms behind evolution, origination, and diversification of taxa. Such dramatic turnovers have therefore played a significant role in the history of life. The 'major five' mass extinctions (Raup and Sepkoski 1982) have received intense attention over the last 25 years, starting with the K-Pg impact hypothesis launched by Luis Alvarez and others in *Science* in 1980. By comparison, few studies have concerned the about sixty smaller-scale bioevents that are scattered throughout the last ca 543 million years (Barnes et al. 1995), although their architecture often resembles that of the large scale mass-extinctions and they therefore may hold important information relevant to events in general.

The Silurian Period is the shortest time period of the entire Phanerozoic (~443-416 Ma) but yields a significant record of ocean-atmosphere-biosphere changes. The period has traditionally been described as an environmentally and faunally stable period in Earth history – a greenhouse period with a moderate latitudinal climate gradient and impoverished marine faunas slowly recovering from the end-Ordovician mass extinction. It has more recently, however, been shown that the Silurian was no quieter than any other time interval during the Phanerozoic Eon, rather the opposite. The results of an immense number of studies, primarily in the fields of taxonomy, biostratigraphy and stable isotope stratigraphy clash fundamentally with the old view. The new studies have unequivocally shown that repeated marine biodiversity crises took place, affecting e.g. graptolites, conodonts, chitinozoans, acritarchs, brachiopods and reefs, and that these turnovers were closely linked to abrupt and significant changes in oceanography and the global carbon cycle. Accordingly, today the Silurian can be regarded as one of the most volatile periods of the entire Phanerozoic when considering the ocean-atmosphere system (Cramer and

Saltzman 2005). In order to understand these changes we need to see beyond the defined stratigraphic boundaries of the Silurian System and treat the time interval in a broader context. This is because the Silurian global events form part of a series of ocean-atmosphere-biosphere changes ranging in time from the Late Ordovician and through the Early Devonian (Fig. 1). In the authors opinion the Silurian global events require fundamental changes in Earth's climate, which are best explained by a series of glaciations. The lack of sedimentary evidence such as tillites from middle and younger Silurian formations has so far hindered the acceptance of this hypothesis.

This is the first thematic publication in more than ten years in which the Silurian global events are treated in close context with the major five extinction events of the Phanerozoic. The previous publication was the result of the IGCP Project 216 'Global Biological Events in Earth History' in which the Silurian bioevents were treated by Kaljo et al. (1995). This chapter calls attention to the tremendous development of Silurian stratigraphy achieved in this last decade and how it generates a fundamentally new view of the Silurian Period as a time interval of repeated global change. The three main events – the Early Silurian Ireviken Event, the Middle Silurian Mulde Event, and the Late Silurian Lau Event – are reviewed with regard to biodiversity changes and how these relate to carbon and oxygen stable isotope evolution and sea-level change. Much of these data presented herein come from the island of Gotland (Sweden; Fig. 2) were a series of stacked carbonate platform generations, ranging in age from the latest Llandovery through the Ludlow, are preserved. Accordingly, this contribution should be seen as 'a view from a carbonate platform' situated on the paleocontinent Baltica in the Silurian tropics, although other areas are referred to as far as possible. In order to stimulate research, facilitate the identification of the Silurian global events elsewhere, and establish a coherent working frame, this paper also serves to present and define the stratigraphical extent of these three events, in terms of first and last appearance datums (FAD's and LAD's), and their expression in shallow-water environments. As the reader will learn, the comparably small-scale Silurian global events share unexpectedly many similarities with the great mass extinctions (Munnecke et al. 2003; Calner 2005a).

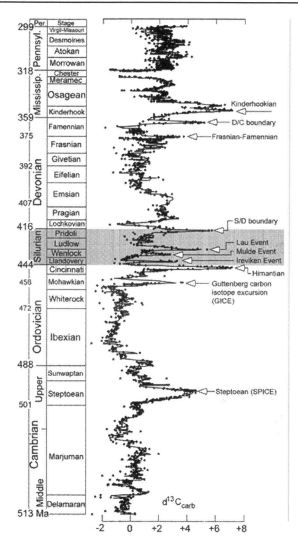

Fig. 1. Paleozoic anomalies in the global carbon cycle based on strata in the Great Basin, USA (modified from Saltzman 2005). The diagram shows a cluster of spectacular anomalies between the Late Ordovician and Late Devonian mass extinction events. These were practically unknown about 15 years ago. Hence, from being considered a 'calm' period in Earth history, the comparably short Silurian Period (grey shaded) has now shown to be one of the most volatile periods during the Phanerozoic, displaying a remarkable history of ocean-atmosphere-biosphere changes that are yet poorly understood. The carbonate platform strata on the island of Gotland, Sweden, form an excellent sedimentary archive for study of the effects of the Ireviken, Mulde, and Lau events in shallow-marine environments

4.2 The Silurian marine scene

During the Silurian the continents were still largely unsettled which means that weathering processes, and the processes of erosion and transport of sediments from the continents to the shallow shelves must have differed substantially from post-Silurian times. The marine realm stood in sharp contrast to the silent land masses and yielded abundant and diverse life forms. The preceding Great Ordovician Biodiversification Event had resulted in a tremendous array of adaptive radiations among marine biota. This in turn resulted in the increased bioturbation rates, the increased number of tiering levels, and the complex trophic structures that came to characterize the Silurian seas some millions of years later. The Silurian faunal provincialism on the other hand was relatively low. This has enabled precise global biostratigraphic correlation based on graptolites that were widespread and common in the black shale facies that is typical for the Silurian. In shallow-water carbonate successions, conodonts fill this purpose. The reef communities seen in the Silurian originated already in the later half of the Ordovician and were only little affected by the terminal Ordovician and Early Silurian glaciations (Copper 2002). Reefs composed of tabulate and rugose corals and stromatoporoids were now prolific on a global scale and, on southern Baltica, these formed extensive reef barriers. Among the most common groups that formed part of the normal marine benthic faunas were brachiopods, trilobites, crinoids, ostracods and bryozoans. A complex pelagic ecosystem is indicated by abundant microfossils such as acritarchs, chitinozoans and graptolites. Pelagic elements with calcitic shells were rare and in contrast to the Mesozoic the carbonate production was almost exclusively benthic. The Silurian global events are far best documented from groups that inhibited the pelagic realm and from the nectobenthic conodonts.

The Silurian starts in the immediate aftermath of a mass extinction that had wide effects on the marine faunas. Based on the number of disappearing taxa the Late Ordovician mass extinction is the second strongest of the major five and wiped out some 50% of the marine taxa. On the other hand it had comparably small ecologic impact on the shallow-marine ecosystems (Droser et al. 2000). The same reef-builders that constructed the Late Ordovician reefs continued to be important in the Silurian reefs. Apart from mass extinction in certain groups, the Silurian global events overall had comparably small taxonomic effects whereas the ecological effects sometimes were comparable to those during the Late Ordovician.

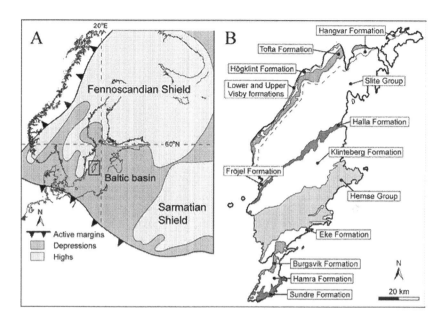

Fig. 2. Maps of the areas discussed herein. A) The Silurian paleogeography of Scandinavia and location of the island of Gotland (within the square) east of the Swedish mainland. The Baltic Basin was a shallow epi- to pericontinetal basin on the southern margin of Baltica and the East European Platform (modified from Baarli et al. 2003). B) Close-up of Gotland showing the geographical distribution of stratigraphic units shown in other figures herein. See Jeppsson et al. (2006) for an updated description of the Gotland stratigraphy

4.3 Discovery of the Silurian global events

Silurian faunal turnovers and extinctions have been known for a long time and the period yield a well documented record of bioevents (Jeppsson 1990; Barnes et al. 1995; Gradstein et al. 2004; Fig. 3). These events have historically been referred to by different names, depending on the group of taxa at hand, and this has from time to time led to some confusion. Along with improved biostratigraphy and stable isotope stratigraphy, however, it is now clear that the various published names sometimes represent the same event. An attempt to name the Silurian events according to their stratigraphic position within the Silurian System was made by Barnes et al. (1995) but this terminology has not been frequently used. The need for a common language for these events has therefore recently been addressed (Calner and Eriksson 2006a). Such a major task is out of the scope of this publication, however, and the names used in this chapter derive from geographical areas on the island of Gotland (Jeppsson 1998). Synonyms are given as far as possible.

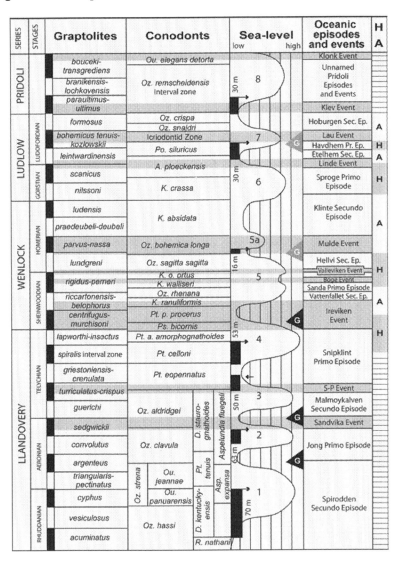

Fig. 3. Stratigraphic framework showing the position of the Silurian bioevents and their close relationship with global sea-level changes. The three most significant events, the Ireviken, Mulde, and Lau events, are discussed in this chapter. The figure is modified from Johnson (2006) with some biostratigraphic data added from Jeppsson (1997a, 2005) and with oceanic event terminology following Jeppsson (1990, 1998). For space reasons, the Lansa Secundo Episode and the Allekvia Primo Episode were excluded from the figure. These are otherwise found between the Boge and Valleviken events. The letters H and A in the right column show the subdivision of the Gotland succession based on the humid (H) and arid (A) climate periods proposed by Bickert et al. (1997). The Early Silurian glaciations (black triangles with a white G) are based on presence of tillites in the Amazon and Paraná basins of Brazil and in the Andean basins of Argentina, Bolivia, and Peru (Grahn and Caputo 1992; Caputo 1998; see also Díaz-Martinez and Grahn 2007) as well as on oxygen isotope data (Azmy et al. 1998). Grey triangles indicate levels for which glaciation has been proposed but still not proved by tillites (Jeppsson and Calner 2003; Lehnert et al. 2007a)

4.3.1 Early discoveries

The perhaps most drastic extinction of the entire Silurian, the near total extinction of the graptolites in the middle Silurian, was known in some detail already at the turn of the last century when British paleontologists identified substantial biodiversity changes among these pelagic colonies. It was not until the end of the 20th century, however, that the scale of this extinction event was fully appreciated among researchers. Similarly, the worldwide crisis among the conodonts in the Early Silurian was indicated already in the stratigraphic charts of Walliser (1964) but fully understood first several decades later when it was documented by Aldridge et al. (1993). The discovery of extinctions among these two particular groups has not been by chance since they by far are the most important for Silurian biostratigraphy and intercontinental correlation – graptolites in fine grained siliciclastic strata formed in the outer shelves and conodonts in carbonate rocks of the inner shelves. Today many more groups have been studied and it has become apparent that the events had a substantial but generally short-lived impact on the marine life and ecosystem. Extinctions and frequency changes among benthic faunas are to date much less well known. This may in part be due to sampling strategies and the simple fact that larger body fossils are not as common per kilogram rock as microfossils.

4.3.2 The theory of Silurian oceanic events

There were no major flood basalt eruptions or extra-terrestrial catastrophes during the Silurian – dramatic processes that are often put forward as causes for mass extinction events. Two impact craters are known from Silurian strata of southern Baltica, representing the Ivar and Kaali bolide impacts (Puura et al. 2000). The Ivar crater formed in the Early Silurian and is situated south of the island of Öland (Sweden). The Kaali crater formed in the Late Silurian and is situated on the island of Saaremaa (Estonia). In the absence of links to these or similar spectacular threats, the causes for the Silurian global events have to date primarily been discussed in terms of normal but abrupt changes in the global climate. The many published reports on Silurian biodiversity changes have therefore in the two last decades focused on oceanographic modelling and the relationships between ocean circulation, primary production and the global carbon cycle. There are today two main competing models for the Silurian climatic shifts. The first model that really opened new pathways for Silurian event stratigraphy was that of Jeppsson (1990). Already in the 1980's Jeppsson had discovered a cyclic stratigraphic pattern in the temporal distribution of conodonts preserved in the carbonate platform rocks of Gotland. The cyclic pattern, which was also paralleled by cyclicity of lithologies, was explained through an oceanic model that described shifts between two stable oceanic-climatic states. These were referred to as Primo Episodes and Secundo Episodes (Jeppsson 1990; Fig. 4). The driving force for the cyclic pattern was changing areas for production of oceanic deep-water over time from high latitudes to low latitudes. Primo Episodes resemble the modern oceanic circulation and climate with low atmospheric CO_2 concentration and cold high latitudes. Given these circumstances, deep oceanic waters were produced through sinking of cold-dense surface waters in the high latitudes. This ventilated the oceans and created upwelling of cold, oxic bottom water in low latitudes, which in turn spurred a high primary production and diverse and abundant shelf faunas. Low latitude climate was humid with common rainfall. The intense weathering produced abundant clay that together with nutrients were shed to shelf seas. The resulting shallowing of the photic zone and excess of nutrients resulted in retreat of the main carbonate producers, reefs, and instead widespread deposition of argillaceous sediments across the shelves. Secundo Episodes were characterized by higher atmospheric CO_2 concentrations and, as a result, the high latitudes were too warm to produce cold-dense surface waters. Instead, the deep oceanic waters formed from sinking of saline-dense surface waters at intermediate latitudes. This generated a stratified ocean with low primary production, less diverse shelf faunas, and formation of black shales in deep water environments. The climate at low latitudes was dry, and as a consequence, low rates of weathering and run-off promoted clear waters and prolific reef growth. At the tipping point between the two stable climatic modes, events occurred due to the obliteration of ocean circulation and collapse of the primary food-web. Considering that the deep-ocean water has a residence time of only a few thousands years the events should be observable 'across a bedding plane'. Since the publication of the model in 1990 new and more effective laboratory methods has been developed in order to dissolve large quantities of limestone without fracturing the tiny conodont elements. This has resulted in the discovery of ten oceanic events scattered through the Llandovery, Wenlock, Ludlow and Pridoli epochs (Jeppsson 1998). Although a great ecological significance of some of these can be questioned, three of the events have proved to be highly significant – by Jeppsson named the Ireviken Event, the Mulde Event, and the Lau Event. These are today known from sections world-wide.

Based mainly on the Gotland succession, the extinction phases of the Silurian events described herein have all been subdivided into smaller segments bounded by *datum points*, that is, stratigraphic levels of disappearance or re-appearance of taxa. Datum points associated with the Ireviken and Mulde events have been identified also in sections remote to the Baltica paleocontinent (Aldridge et al. 1993; Jeppsson 1997a; Porębska et al. 2004; Pittau et al. 2006) exluding Signor-Lipps effects and instead suggesting a global significance of the stepwise biodiversity changes. The stepwise pattern of extinction and origination, which tentatively has been attributed to Milankovitch cyclicity (Jeppsson 1997a), needs to be further evaluated. An important parallel result of Jeppsson's work is a conodont-based biostratigraphic framework that is useful for much of the Silurian shallow shelf strata on a global scale and which can be used as a complement to the graptolite biostratigraphy.

4.3.3 The increased use of stable isotope stratigraphy

There are no doubts that the global importance of the Ireviken, Mulde, and Lau events has been confirmed and received increased attention because of the numerous carbon and oxygen stable isotope studies reported from around the globe. Anomalies in the $\delta^{13}C$ composition in calcite and organic carbon show that these events all are associated with profound positive carbon isotope excursions, reflecting perturbations in the global carbon cycle and thereby the global climate (Talent et al. 1993; Samtleben et al. 1996; 2000; Kaljo et al. 1997, 2003, 2006; Saltzman 2001; Munnecke et al. 2003; Cramer and Saltzman 2005; Calner et al. 2006a; Cramer et al. 2006; Lenz et al. 2006).

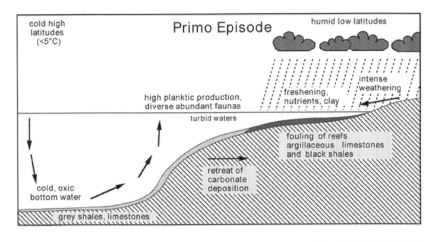

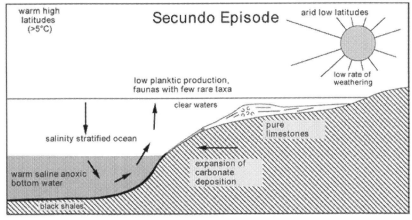

Fig. 4. The oceanic-climatic model of Jeppsson (1990). See text for explanation

Although such changes were predicted by Jeppsson (1990) they were not fully explained by the model – at the time of publication, Jeppsson's model was one of the first efforts to produce a global oceanic model based on the temporal cyclicity of microfossils and lithology. Subsequent studies showed that primo episodes had generally low carbon and oxygen isotopic values whereas secundo episodes had high such values (Samtleben et al. 1996; Bickert et al. 1997). By including the complex arrays of biological and physical fractionation of the carbon and oxygen stable isotopes and oceanography, the Jeppsson model has therefore been scrutinized and new, competing models have been developed. Bickert et al. (1997) argue that the oxygen isotopic shifts reflect varying rates of fresh water input (i.e. salinity) rather than temperature changes. Accordingly, they explained the climatic shifts as caused by changes in the global evaporation-precipitation rates and rate of continental run-off. They modified the Jeppsson model with respect to ocean circulation and stable isotope geochemistry and referred to shifts between humid periods with high terrigenous input and an estuarine circulation (H-periods) and arid periods with little continental run-off and anti-estuarine circulation (A-periods; Fig. 5). The H-periods and A-periods are longer than the primo and secundo episodes and therefore exclude some of the biodiversity trends that are of importance for the original model.

With time more emphasize has been put on explaining the profound carbon isotope excursions. One long-standing problem has been to explain where carbon was buried during the carbon isotope excursions. This may be a matter of correlation problems and the fact that deep oceanic sediments from the middle Paleozoic is lacking due to continental drift and consummation of oceanic plates in subduction zones. This lack of corresponding organic-rich shales led Cramer and Saltzman (2007) to develop the two prevailing oceanic models further. A correspondence between the isotope excursion and organic-rich shales has, however, been demonstrated for the Mulde Event (Porębska et al. 2004; Calner et al. 2006a).

4.4 The Early Silurian Ireviken Event

Extinctions close to the Llandovery-Wenlock boundary have been known for a long time and during this event the most profound Silurian extinction of conodonts took place. The corresponding event interval, about thirteen metres thick, outcrops along a distance of 60 kilometres on north-western Gotland.

4.4.1 Stratigraphic position

The Ireviken Event starts close below the Llandovery-Wenlock epoch boundary. In shelly sequences this corresponds to the zonal boundary between the *Pterospathodus amorphognathoides* and the Lower *Pseudooneotodus bicornis* conodont zones (Fig. 6). The end of the event in shelly sequences is put at the top of the Lower *Kockelella ranuliformis* conodont Zone (Jeppsson 1997b). In terms of graptolite biostratigraphy this corresponds to the upper part of the *Cyrtograptus lapworthi* graptolite Zone (Loydell et al. 1998, 2003).

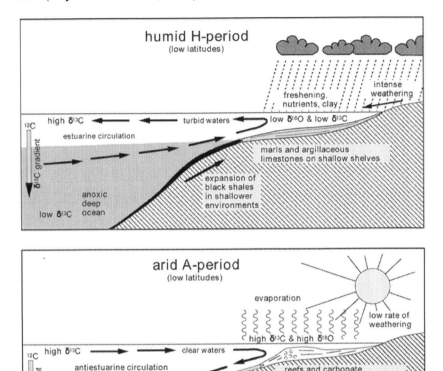

Fig. 5. The oceanic-climatic model of Bickert et al. (1997)

In chitinozoan biostratigraphy the start of the event is close to the base of the *Margachitina margaritana* Zone as defined at the Llandovery-Wenlock Global Stratotype Section and Point (GSSP) at the Hughley Brook section, United Kingdom (Mullins and Aldridge 2004). The stratigraphic range of the Ireviken Event has been calculated as corresponding to ca 200,000 years with faunal changes recorded at eleven different datum points (Jeppsson 1997a, 1998; Munnecke et al. 2003).

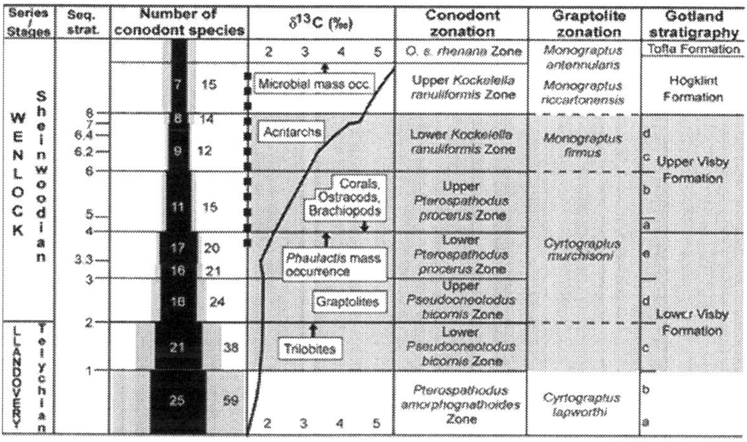

Silurian global events 33

Fig. 6. The stratigraphic range of the Ireviken Event (grey area), the position of datum points (shown in the sequence stratigraphy column), and the diversity changes in conodonts (grey – globally, black – Gotland). *Phaulactis* is a large rugose coral that literally covered the sea-floor during a brief, early part of the event. The figure is modified from Munnecke et al. (2003) with conodont data from Jeppsson (1997b). Graptolite zones are based on Loydell et al. (1998, 2003). Note that extinctions among acritarchs occurred throughout the event although the majority (>80%) of these took place in the uppermost ca four metres of the Upper Visby Formation (Gelsthorpe 2004). Dashed vertical line indicates proposed glaciation by Brand et al. (2006)

4.4.2 Groups affected

The Ireviken Event is characterized by a survival ratio among conodonts of merely twelve out of about sixty globally known species (Aldridge et al. 1993; Jeppsson 1997a). Conodont diversity never recovered during the remaining Silurian and stayed at about twenty taxa. Also the graptolites went through a profound extinction reducing the global fauna by 80% (the *Cyrtograptus murchisoni* Event of Melchin et al. 1998; Noble et al. 2006). Chitinozoans display a significant turnover in assemblages and decrease in diversity (cf. Laufeld 1974; Hints et al. 2006). The majority of the chitinozoan taxa became extinct during the event (Nestor et al. 2002). The acritarchs, most likely forming a significant division of the ocean's primary producers, went through a major turnover during the event; forty-four species became extinct and fifty-four species originated with most of the extinctions at the end of the event (Gelsthorpe 2004). Comparably little is known about the vagrant benthos although they were clearly affected. For example, based on studies of scolecodonts (polychaete jaws), more that 20% of the polychaete fauna went extinct (Eriksson 2006). There are good indications also for a substantial decline in trilobite taxa close to the Llandovery-Wenlock boundary, at the 2nd datum of the event. Such data are based on literature and include no less than a 50% decline of trilobites on southern Baltica (Ramsköld 1985). Extinctions took place also among the brachiopods. Although the amount of brachiopod taxa that was eliminated was small, these extinctions were notable in terms of shelly biomass and because of the decreased abundance of several groups that previously was important (Kaljo et al. 1995). A faunal turnover in Lingulid brachiopods have been documented from Australia Valentine et al. (2003).

4.4.3 Stable isotopes

Selected areas and peak values for the carbon isotope excursion associated with the Ireviken Event: The excursion shows $\delta^{13}C$ peak values of ca 3‰ in Nevada and ca 4‰ in Oklahoma (Saltzman 2001), ca 5‰ in Sweden (Munnecke et al. 2003); 6.6‰ in Norway (Kaljo et al. 2004), ca 2.6‰ in Arctic Canada (Noble et al. 2005), ca 5.5‰ in New York State and Ontario (Brand et al. 2006).

Fig. 7. The Silurian global paleogeography was dominated by the vast Gondwanan continent, which covered much of the southern circumpolar areas. Laurentia, Baltica, and Avalonia assembled in the large Laurussian contintent at equatorial latitudes (modified from Cocks and Torsvik 2002). The figure shows locations where previous studies have presented firm evidence for anomalies either in stable isotopes, biodiversity, or facies during the Ireviken Event. Stars with white dot indicate that the cited study does not include stable isotope data. Note that areas outside Laurussia are poorly represented. AC – Arctic Canada (Noble et al. 2005), AL – Alaska (see Jeppsson 1997a), AU – Austria (Wenzel 1997), EB – East Baltic Area (Kaljo et al. 1997, 1998), GB – Great Britain (Jeppsson 1997a; Munnecke et al. 2003), IO – Iowa (Cramer and Saltzman 2005), NE – Nevada (Saltzman et al. 2001), NO – Norway (Kaljo et al. 2004), NSW – New South Wales (Talent et al. 1993), NY – New York and Ontario (Brand et al. 2006), NWT – North West Territories (see Jeppsson 1997a), OH – Ohio (Cramer and Saltzman 2005), OK – Oklahoma (Saltzman 2001), QU – Quebec (Azmy et al. 1998), SW – Sweden (Munnecke et al. 2003), TE – Tennessee (Cramer and Saltzman 2005)

4.4.4 Sedimentary changes and sea-level

Johnson (2006) noted that the Ireviken Event appears to start somewhat after the onset of global sea-level fall and terminate before the maximum lowstand, meaning that the event took place entirely during a regression. In the East Baltic area a hiatus and sequence boundary has been proposed at the level of Datum 2 (Nestor et al. 2002; Loydell et al. 2003), i.e. within the earliest part of the event.

4.5 The Middle Silurian Mulde Event

The Middle Silurian Mulde Event refers to a faunal overturn and crisis at the top of the Lower Homerian stage. A spectacular mass extinction among the graptolites characterizes the event. The mass mortality is locally expressed by extremely dense bedding plane accumulations of a few single survivor species, sometimes literally blackening the bedding planes (Lenz et al. 2006). Contemporaneous formation of organic-rich shale, alum shale, and not least the formation of the so-called 'boundary coal seam' in deep-water environments shows that anoxia were widespread during the peak of the crisis (Jaeger 1991). The event was first discovered in shale basins by graptolite researchers and extinctions related to this event have alternatively been referred to as 'die große Krise' (Jaeger 1991), C_1 (Urbanek 1970), the *lundgreni* Event (Koren' 1991), or as the *lundgreni*

Extinction Event (LEE; Lenz et al. 2006). The changes in biota associated with this event were recognized early and were one of the reasons for the original placement of the Wenlock-Ludlow boundary at this level.

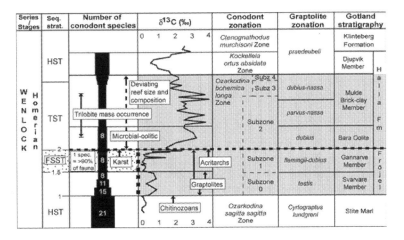

Fig. 8. The stratigraphic range of the Mulde Event (grey area), the position of datum points, and the diversity changes in conodonts (Gotland only). Conodont stratigraphy from Calner and Jeppsson (2003) and graptolite stratigraphy from Porębska et al. (2004). Other data from Pittau et al. (2006; chitinozoans), Porębska et al. (2004; graptolites), Calner et al. (2006a; acritarchs and trilobites) and Calner (2005a; reefs). Sequence stratigraphy from Calner (1999) and Calner et al. (2006a). Stippled pattern marks influx of sand-sized siliciclastics into this carbonate basin. Dashed vertical line indicates proposed glaciation by Jeppsson and Calner (2003). Dashed horizontal line marks the level for carbonate platform termination

4.5.1 Stratigraphic position

The Mulde Event starts at the LAD of the graptolite *Cyrtograptus lundgreni* in shale successions and at the LAD of the conodont *Ozarkodina sagitta sagitta* in carbonate successions (Fig. 8). The end of the event is placed at the FAD of the graptolite *Colonugraptus praedeubeli* in shale successions, which nearly corresponds to the FAD of the conodont *Ctenognathodus murchisoni* in carbonate successions. Extinctions took place at three datums (Jeppsson and Calner 2003; Porębska et al. 2004).

4.5.2 Groups affected

This event has often been put forward as primarily affecting pelagic groups. Indeed, the graptolite extinctions are the most significant effects of the event, eliminating some 95% of the species globally – a lower Homerian fauna of about 50 species was reduced to two or three in the basal upper Homerian (Melchin et al. 1998; Lenz and Kozłowska-Dawidziuk 2001; Porębska et al. 2004; Lenz and Kozłowska 2006). The diverse and specialized pre-event graptolite fauna consisted of cyrtograptids, streptograptids, testograptids, pristiograptids, monoclimacids, paraplectograptids, retiolitids, sokolovograptids and cometograptids (Porębska et al. 2004). The graptolites went extinct in three successive steps leaving behind extremely poor faunas in the *dubius*, *parvus-nassa*, and *dubius–nassa* zones (*sensu* Porębska et al. 2004). Studies from Arctic Canada show that also the planktonic radiolarians were severely affected during the event. The pre-extinction strata yield a diverse fauna of twenty-eight species. Only two of these species survive into the recovery phase of the event (Lenz et al. 2006). Acritarchs were reduced by 50% early during the event (Porębska et al. 2004). This group shows a short-lived but remarkable increase in abundance just prior to the onset of the event both on Gotland (Calner et al. 2006a) and in Poland (Porębska et al. 2004). Studies of deep-water black shales on Sardinia have shown that chitinozoans went through two extinction phases (Pittau et al. 2006). Mass mortality among nautiloid cephalopods was noted by Lenz et al. (2006) near the base of the *dubius-nassa* zone. Moderate extinctions took place also among conodonts (Jeppsson et al. 1995; Jeppsson and Calner 2003) and the first of two clear Silurian extinction events among fish took place during this event (Märss 1992).

There is no good evidence for substantial taxonomic loss among benthic biota although there are certainly community-scale and biodiversity changes related to the interval. The appearance of the graptolite Lilliput taxa *Pristiograptus dubius parvus* immediately after the extinctions marks the lower boundary of the *parvus-nassa* Zone and the earliest survival phase of the graptolite extinction (Porębska et al. 2004). This level is associated with a distinct change in the benthic faunas of the Prague basin were new brachiopod and trilobite communities appear (Kaljo et al. 1995). Similarly, a short-lived mass-occurrence of the trilobite *Odontopleura ovata* has been documented widely during the extinction and earliest recovery phase of the event (Calner et al. 2006b). Based on literature it is clear that substantial changes took place in various benthic groups during this time interval (Hede 1921).

The stratigraphic interval for the graptolite extinctions has more recently been thoroughly investigated with respect to sedimentology change in the carbonate platform strata of Gotland (Calner 1999, 2002; Calner and Säll 1999; Calner and Jeppsson 2003; Calner et al. 2006a). These studies show that the extinction among pelagic taxa correlates perfectly with termination of carbonate platforms, karst formation, and development of oolites and microbial carbonates in carbonate platforms. The extensive carbonate platforms in the cratonic interiors of the American Midwest show similar abrupt changes in carbonate production (e.g. in Indiana and Kentucky).

4.5.3 Stable isotopes

Selected areas and peak values for the carbon isotope excursion associated with the Mulde Event: The carbon isotope excursion associated with the Mulde Event has been recognized globally (Fig. 9). The $\delta^{13}C$ peak values reach 4.6‰ in the East Baltic Area (Kaljo et al. 1997), 3.8‰ on Gotland (Calner et al. 2006a), ca 2.5‰ in Nevada (Cramer et al. 2006), ca 2.8‰ in Tennessee (Cramer et al. 2006).

4.5.4 Sedimentary changes and sea-level

This event is related to pronounced facies shifts in Laurentia, Baltica and peri-Gondwana, independent of basin type: contemporaneous facies anomalies are known from rapidly subsiding rift basins (Kaljo et al. 1995), from deep shale basins (Porębska et al. 2004) as well as from shallow intercontinental carbonate platforms (Calner and Jeppsson 2003) and off-platform slope and basin settings (Calner et al. 2006a; Lenz et al. 2006). In shaly successions on peri-Gondwana as well as on Baltica, the event is associated formation of phophorites (Pittau et al. 2006; Jaeger 1991). On Gotland, sea-level fall during the earliest part of the event (*flemingii-dubius* Zone) is manifested by influx of siliciclastic material and formation of epikarst and a rocky shoreline with an erosional relief of at least 16 m (Calner and Säll 1999; Calner 2002). In this area, the maximum sea-level lowstand (the karst surface) is at or very close to Datum 2 of the event. The sea-level lowstand surface is overlain by a thin oolitic unit that can be followed across Gotland and which reappears at the same level in the East Baltic Area more than 200 km away. The oolite formed during the earliest part of a marked transgression that continued through the *dubius, parvus-nassa,* and *dubius-nassa* zones. This transgression gave rise to famous stratigraphic units such as the Mulde Formation on Gotland (now termed the Mulde Brick-clay Member) and the Waldron Shale across south and central USA. Equivalents to these shaly units are found also in Arctic Canada (cf. Lenz et al. 2006). The time period of extremely low diversity and carbon isotope excursion correlates with increased deposition of organic carbon in the oceans (Porębska et al. 2004; Calner et al. 2006a; Lenz et al. 2006). The precise relationship between extinctions, sea-level change and carbon isotope excursion in carbonate platforms has been demonstrated by Calner et al. (2006a).

Middle Silurian Mulde Event

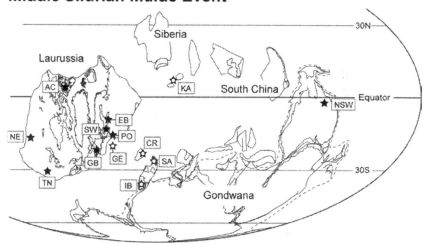

Fig. 9. The figure shows locations where previous studies have presented firm evidence for anomalies either in stable isotopes, biodiversity, or facies during the Mulde Event. Stars with white dot indicate that the cited study does not include stable isotope data. AC – Arctic Canada (Lenz et al. 2006), CR – Czech Republic (Kaljo et al. 1995; Kozłowska-Dawidziuk et al. 2001), EB – East Baltic Area (Kaljo et al. 1997, 2003), GB – Great Britain (Corfield et al. 1992), GE – Germany (Jaeger 1991), IB – Iberian Peninsula (Gutiérrez-Marco et al. 1996), KA – Kazakhstan (Koren'1991), NE – Nevada (Cramer et al. 2006), NSW – New South Wales (Rickards et al. 1995), PO – Poland (Porębska et al. 2004), SA – Sardinia (Pittau et al. 2006), SW – Sweden (Calner et al. 2006a), TN – Tennessee (Cramer et al. 2006).

4.6 The Late Silurian Lau Event

The Lau Event is associated with the most spectacular positive carbon isotope excursion of the entire Phanerozoic and its vast magnitude is difficult to explain by normal processes. The associated faunal crisis has for a long time been recognized as an important event in the late Ludfordian and is known to have affected pelagic as well as benthic faunas. Substantial ecosystem changes are known from carbonate platforms were the composition of reef-builders change during the event interval. Extinctions related to this event has alternatively been termed the Pentamerid Event (Talent et al. 1993), Podoliensis extinction event (Koren' 1993), the Cardiola Event (Schönlaub 1986), or the Kozlowskii Event, including two phases of extinction (cf. Manda and Kříž 2006). Extinctions took place at five datums.

4.6.1 Stratigraphic position

In carbonate successions the event starts just before the LAD of the conodont *Polygnathoides siluricus* and ends near the FAD of the conodont *Ozarkodina snajdri* (Jeppsson and Aldridge 2000; Fig. 10). In shaly successions the start of the event has been identified very close to the LAD of the graptolite *Neocucullograptus kozlowskii* (Lehnert et al. 2007b). The pronounced carbon isotope excursion and the LAD of *P. siluricus* in the earliest part of the crisis have facilitated precise identification of the event in widely spaced sections, e.g. on Gotland (Jeppsson and Aldridge 2000), in the Mušlovka section of the peri-Gondwanan Prague Basin (Lehnert et al. 2003; 2007b), and in the Broken River area, Queensland (Talent et al. 1993).

4.6.2 Groups affected

The Lau Event caused considerable extinctions and changes in community structures (Jeppsson and Aldridge 2000). Among conodonts, no platform-equipped taxon survived, and disaster conodont faunas dominated by a single taxon developed during the most severe part of the event. Out of twenty-three conodont species present in pre-extinction strata, seventeen had disappeared by the end of the event (Jeppsson and Aldridge 2000). Although the data on conodonts are by far the most detailed, it is evident from the literature that several other taxonomic groups were also affected by this event. Throughout the continent Baltica, marine vertebrates went through their second distinct Silurian extinction event, replacing much of the pre-event fauna (Märss 1992; Kaljo et al. 1995). The event has for long times been documented also from shale basins, reducing 70% of the pelagic graptolites (Koren' 1993; Storch 1995; Melchin et al. 1998). Among benthos, the event records a dramatic reduction in the diversity of pentamerid brachiopods (Talent et al. 1993) and a conspicuous reorganization of brachiopod faunas in shallow seas (Gustavsson et al. 2005). The appearance of monospecific brachiopod communities appears on different continents, e.g. *Dayia navicula* on Baltica (Jeppsson et al. 2007) and *Dayia minor* in Bohemica (Lehnert et al. 2007b). Also the bivalves went through extinctions and substantial faunal reorganisations (Schönlaub 1986). Many other groups for which ranges of species has been documented show distinct changes during the event. Among these are chitinozoans (Laufeld 1974), ostracodes (Martinsson 1967), polychaetes (Bergman 1989), and acritarchs (Le Hérissé 1989).

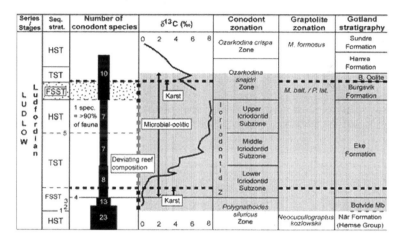

Fig. 10. The stratigraphic range of the Lau Event (grey area), the position of datum points, and the diversity changes in conodonts (grey – globally, black – Gotland). Conodont stratigraphy and biodiversity data from Jeppsson (2005) and Jeppsson et al. (2007), respectively. Reef data from Calner (2005a). Stable isotope data from (Calner and Eriksson 2006b) and from Samtleben et al. (2000). Sequence stratigraphy from Eriksson and Calner (in press). Stippled pattern marks influx of sand-sized siliciclastics into the basin. Dashed vertical line indicates proposed glaciation by Lehnert et al. (2007a) and Eriksson and Calner (in press). The grey section of this line marks a minor transgressive event. Dashed horizontal lines mark two successive levels for carbonate platform termination

4.6.3 Stable isotopes

Selected areas and peak values for the carbon isotope excursion associated with the Lau Event are shown in Figure 11: This event is associated with one of the largest positive carbon isotope excursions of the entire Phanerozoic. The excursion shows $\delta^{13}C$ peak values of up to 12‰ in Australia (Andrew et al. 1994), ca 5‰ in the East Baltic Area (Kaljo et al.

1997), 11.2‰ in southernmost Sweden (Wigforss-Lange 1999), 8.8‰ on Gotland, Sweden (Samtleben et al. 2000), ca 4‰ in Oklahoma and Nevada (Saltzman 2001), and ca 8‰ in Bohemia (Lehnert et al. 2007b).

Late Silurian Lau Event

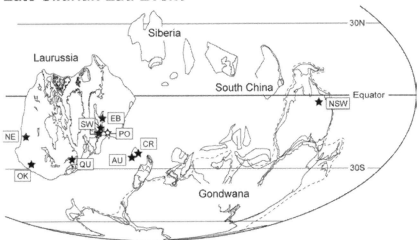

Fig. 11. The figure shows locations where previous studies have presented firm evidence for anomalies either in stable isotopes, biodiversity, or facies during the Lau Event. Stars with white dot indicate that the cited study does not include stable isotope data. AU – Austria (Schönlaub 1986; Wenzel 1997); CR – Czech Republic (Lehnert et al. 2003, 2007b), EB – East Baltic Area (Kaljo et al. 2003), NE – Nevada (Saltzman 2001), NSW – New South Wales (Talent et al. 1993), OK – Oklahoma (Saltzman 2001), PO – Poland (Kozłowski 2003), SW – Sweden (Samtleben et al. 2000; Calner and Eriksson 2006b)

4.6.4 Sedimentary changes and sea-level

The most pronounced sedimentary changes in the Late Silurian of southern Baltica are associated with the Lau Event, from Poland in the southeast to Gotland in the north and further to the province of Skåne (Sweden) in the southwest. Influx of terrigenous sands and formation of epikarst show that the facies evolution was forced by a substantial sea-level drop (cf. Kozłowski 2003; Eriksson and Calner, in press). A contemporaneous sea-level drop is well documented also from the peri-Gondwana Prague basin (Lehnert et al. 2007a-b) and from Australia (Jeppsson et al. 2007). Based on the succession of Gotland there were two regressions, which started in the uppermost *Polygnathoides siluricus* Zone and lower *Ozarkodina snajdri* Zone, respectively. The event coincides with a microbial resurgence and formation of oolites (Calner 2005b).

4.7 Structure of the Silurian global events

This section aims to give a brief overview of the general structure of the Silurian global events and their similarities and differences with some of the major five mass extinction events with regard to biodiversity, stable isotopes and sea-level change. The temporal development for each event is shown in Figures 6, 8, 10.

It has commonly been assumed that the Silurian global events had little effect on vagrant and sessile benthic faunas. This is not true for the low latitudes were diverse marine ecosystems (carbonate platforms) thrived through much of the Silurian. As discussed below, these ecosystems changed markedly during the events due to the interaction of climate, sea-level and biotic changes, but re-established contemporaneously with the recovery of faunas and declining carbon isotope anomalies. Hence, a rather significant ecological impact of the Silurian global events on shallow marine environments may have been decoupled from a less drastic taxonomic impact (cf. Droser et al. 1997, 2000). This section first discusses a few major changes in carbonate platforms during the Silurian events. Thereafter the events are summarized with regard to their temporal and spatial development, biodiversity, carbon isotopes and sea-level changes.

4.7.1 The Silurian events and carbonate platforms

Todays carbonate platforms – such as the Bahamas platform or the Great Barrier Reef – are major marine ecosystems that respond to a wide variety of changes in the contemporaneous oceans. They have a life cycle and a carbonate production rate that depends on the interaction of climate, relative sea-level, and biotic factors such as diversification rate. For this reason, environmental changes in coastal marine waters are reflected in the type of carbonate production and thus in the sediment composition of carbonate platforms. The changes are generally profound at times of major global crises and a number of anomalous carbonate facies can be directly related to mass extinction events. As opposed to the normal marine shelly faunas of pre- and post-extinction strata the anomalous carbonate facies is often microbially mediated and associated with a substantial increase in calcifying cyanobacteria, leading to mass occurrences of oncoids and sessile stromatolites in normal marine environments. Other anomalous facies common to extinction events are flat-pebble conglomerates (Sepkoski 1982) and wrinkle structures (Hagadorn and Bottjer 1997; Pruss et al. 2004). This type of signals of ecologic turnover are known from the Late Ordovician (Sheehan and Harris 2004), the Late Devonian (Whalen et al. 2002), and from the Permian-Triassic (Schubert and Bottjer 1992) mass extinctions. There are two prevailing ways of interpreting the overrepresentation of microbially mediated facies during mass extinctions. The first is biological and suggests that the increase of cyanobacteria is due to lowered grazing pressure by vagrant benthos (Hagadorn and Bottjer 1999). The other interpretation favours changing ocean chemistry and relates the same increase in abundance to increased carbonate saturation states in the oceans (Riding and Liang 2005).

Fig. 12. Carbonate platforms are major marine ecosystems which inhabitants respond to environmental changes in the contemporaneous oceans. Fossil carbonate platforms, such as those on the island of Gotland, are important archives from which biotic changes and anomalies in the carbon cycle, as well as physical and chemical oceanographic changes, can be measured and gauged in the same stratigraphic section. Based on such integrated analysis of these strata, the Ireviken, Mulde, and Lau events are all associated with profound environmental changes in the ecology of coastal environments. The photograph shows three major reefs of the Högklint Formation on north-western Gotland

The several stacked carbonate platform generations that form the bedrock of Gotland (Fig. 12) have been highly useful for integrated studies of the Silurian global events. Recent studies have shown that the Mulde and Lau events are associated with ecosystem changes in low-latitude shallow seas that are similar to those during the mass extinctions discussed above. This is expressed as short-lived but prominent lapses in biotic carbonate production during which microbial carbonates regionally were important. A microbial resurgence, and the formation of microbial-ooliticstromatolitic sedimentary units, has been demonstrated in strata formed during the Lau Event when stromatolites and wrinkle structures increases substantially in abundance or appear during or shortly after the extinction phase (Calner 2005a, b, c). These are the first wrinkle structures documented from strata formed during a smaller extinction event and the first reported from the Silurian. A similar but less conspicuous microbial resurgence occurred also during the extinction phase of the Mulde Event and after the Ireviken Event (Munnecke et al. 2003; Calner 2005a). The oolites that developed during the Mulde and Lau events are of particular interest since they have important analogues during the end-Permian mass extinction when these non-biological carbonates dominated along the shores of the Tethys Ocean (Groves and Calner 2004). The oolites formed during or immediately after the extinction events and superimpose terminated carbonate platforms with abundant skeletal carbonate production and large reefs. Widespread oolites are therefore lithologic analogues to low-diversity, post-extinction disaster faunas common to the majority of Phanerozoic events – a sign of ecological stress in the Silurian shallow seas. This may be true for Phanerozoic extinction events in general and for the widespread Holocene interglacial oolites.

The reefs that follow immediately after the Mulde and Lau events on Gotland are small and have a deviating composition as compared to the pre-event reefs. The main reef-builders, the stromatoporoids, decrease markedly in diversity and size and is to some extent replaced by dendroid or encrusting tabulates and bryozoans (Calner 2005a). The significance of these changes is not yet revealed.

4.7.2 Temporal and spatial development and biodiversity

To be regarded a mass extinction an event should be (1) 'confined to a short interval of geological time, it should (2) affect a wide variety of clades occupying a wide spectrum of habitats and (3) it should eradicate a high proportion of species' (Brenchley and Harper 1998, p. 322; numbers added by the current author). These three prerequisites are only partly fulfilled by the Silurian global events, the minor amount of taxonomic loss during the Silurian global events being the main difference.

1. The Silurian global events are characterized by a short duration. Calculations based on stepwise conodont extinctions suggest that the initial phase of extinction are within a range of a few 10,000's to about 200,000 years, and is followed by recovery phases that are of comparable length or slightly longer (Jeppsson 1997a; Jeppsson and Calner 2003). This is clearly an advantage in Silurian event studies since it means that the associated temporal changes in fauna and facies generally can be studied within restricted outcrop areas or in relatively short drillcores, even in stratigraphically expanded successions.

2. The Silurian global events affected a wide range of clades across a wide range of habitats, from shallow-water coastal environments to the oceanic realm. This has recently been demonstrated for the Mulde Event in a study based on sections in Sweden, Poland and Germany. A high-resolution correlation between epicontinental carbonate platform strata (Sweden) and deep-water strata formed off the Silurian cratonic margin (Poland and Germany) has shown that synchronous biodiversity changes took in planktonic groups such as graptolites and acritarchs, nektonic taxa such as conodonts, and trilobites, the latter representing vagrant benthos. An associated sea-level drop simultaneously terminated shallow water carbonate platforms. Hence, the event affected the entire marine realm, from shoreline to deep basin, and the entire water column, from sea-floors well below the storm wave base to the upper parts of the oceans where planktonic taxa thrived (Calner et al. 2006a).

3. The Silurian biodiversity changes and extinctions have numerically hitherto mainly been adequately demonstrated from microfossil assemblages representing nectonic or planktonic life modes, in particular the conodonts (e.g. Jeppsson 1990, 1997a; Jeppsson et al. 1995; Jeppsson and Aldridge 2000), graptolites (e.g. Melchin et al. 1998; Lenz and Kozłowska-Dawidziuk 2001), and acritarchs (e.g. Porębska et al. 2004; Calner et al. 2006a). These groups are well studied and extinction of 80% and 95% of the global fauna has been recorded for conodonts in the earliest Wenlock and graptolites in the Late Wenlock, respectively. By comparison, only little numerical data exist for contemporaneous benthos although major biodiversity changes are indicated in the literature.

Ecological changes on a species level are common during the Silurian global events. For example, the post-extinction dwarfing of species – the 'Lilliput effect' – that in the last decade has been recognized in the aftermath of all the major five mass extinction events, and which appears to be common to all biological events independent of scale and forcing mechanisms (Twitchett 2001), was first documented among Silurian graptolites (Urbanek 1993). A good example is graptolite *Pristiograptus dubius parvus* in the immediate aftermath of the Mulde Event. Lilliput taxa have also been documented among scolecodonts (Eriksson et al. 2004).

4.7.3 Carbon isotope stratigraphy

The Ireviken, Mulde and Lau events were all associated with changes in the global carbon cycle that were of equal or even greater magnitude than those related to the end-Ordovician and end-Devonian mass extinctions. Hence, the perturbations in the ocean-atmosphere system during the intervening Silurian events are fully comparable to those during the major five mass extinction events. This is an important observation since it clearly indicates that the series of comparable perturbations in the ocean-atmosphere system during the middle Paleozoic (Fig. 1) was associated with widely different biodiversity and ecosystem changes.

Stable carbon isotope ($\delta^{13}C$) stratigraphy of marine carbonates and organic material has to date pin-pointed the Ireviken, Mulde and Lau events on remote paleocontinents (Figs. 7, 9, 11). Based on the Gotland succession, the temporal development and architecture of the positive carbon isotope excursions are broadly speaking similar between the different events: The onset of the positive carbon isotope excursion is generally close to or succeeds the first extinction datum. Following an initial short period with slowly increasing values from a base-level they suddenly increases abruptly with the mot rapid rate of increase of $\delta^{13}C$ values just before or within the interval with the lowest diversity. Peak values are reached within the low-diversity interval and the curve thereafter describes a protracted decline back to a new base-level during the recovery phase (Figs. 6, 8, 10). During the Mulde and Lau events, the most rapid rate of increase of $\delta^{13}C$ values straddles epikarstic surfaces in carbonate platforms. Accordingly, the $\delta^{13}C$ excursions started during the termination phase of carbonate platforms and escalated during the subsequent drowning of that platform and the initiation phase of a new platform (cf. Cramer and Saltzman 2007). This may also be true for the Ireviken Event if the interpretation of a hiatus (lowstand) at the Llandovery-Wenlock boundary is correct (Loydell et al. 2003).

4.7.4 Sea-level change and sedimentary facies

The majority of the Phanerozoic extinction events are associated with some degree of sea-level change (Hallam and Wignall 1999), although the extinctions not necessarily are caused by the sea-level change. It is strikingly that the Silurian events are in the majority of investigated places associated with pronounced facies shifts, deviating, or even peculiar, sedimentary facies (Kaljo et al. 1995). Some of these facies are difficult to explain by sea-level change and may be the result of more complex environmental-biological interactions. Based on a excellent set of biological, geochemical and physical data, assembled under more than two decades, seven out of the eight Silurian oceanic events (*sensu* Jeppsson 1998) show good correlation with global sea-level lowstands (Johnson 2006; Fig. 3). Accordingly, brief hiatuses and epikarst are associated with some of the events in cratonic interiors. Based on such settings, the somewhat complex relationship between sea-level change and extinction datums during the Silurian global events are illustrated in Figures 6, 8, and 10. For the Mulde and Lau events these diagrams show that extinctions primarily are related to regressions and lowstands (the falling stage systems tract) and that the low-diversity catastrophy faunas existed during the early transgression and possibly early highstand. The sequence stratigraphy during the Ireviken Event has not yet been analyzed in detail although a fairly rapid relative shallowing from below the storm wave base to above the fair weather wave base is obvious from facies.

The causes for the sea-level changes during the Silurian events have not been unequivocally demonstrated although the associated carbon cycle anomalies suggest that they are related to climate change. Facies evidence for shallowing always predates the most rapid rate of increase of $\delta^{13}C$ values. Nevertheless, considering the known causes for rapid sea-level change, glaciations are the most plausible explanation for the development during the Ireviken, Mulde and Lau events. This has support also from stable oxygen isotopes measured on conodont apatite (Lehnert et al. 2007a). The reworked tillites in the Amazon and Paraná basins of Brazil and in the Andean basins of Argentina, Bolivia, and Peru (Grahn and Caputo 1992; Caputo 1998) show that the Late Ordovician cool climate persisted well into the Early Silurian. The fact that extensive landmasses covered the southern polar areas did not change through the Wenlock and Ludlow (Cocks and Torsvik 2002) and it is therefore suggested that tillites should be questioned as sole evidence for glaciation with regard to their poor preservation potential. It should although be emphasized that later Silurian glaciations may have been of less geographical extent and therefore the corresponding sea-level falls of less magnitude. The assembly of Laurussia and the Caledonian Orogeny overlaps in time with the distribution of carbon cycle anomalies shown in Figure 1. These major tectonic events created major mountain chains in the tropics and subtropics. With regard to this setting, the increased weathering rates is one argument for increased consumption of atmospheric CO_2 and cooling of the climate from the Late Ordovician through the entire Silurian.

4.7.5 Taxonomic vs ecologic events

The scale of mass extinctions is usually ranked by the severity of taxonomic loss (Raup and Sepkoski 1982). Ecologic changes associated with bioevents are more difficult to quantify and attempts to develop a systematic method for identifying *paleoecological levels* have been made (Droser et al. 1997, 2000). They included microbial resurgence in normal marine environments as a characteristic signal of their second-level paleoecological changes during mass extinctions. Hence, from an ecological point of view, the Silurian global events stand as remarkable miniatures of the more devastating extinction events of the Phanerozoic (Munnecke et al. 2003; Calner 2005a).

In conclusion, apart from the obvious differences in taxonomic loss, the Silurian events are in several aspects comparable to the five monumental crises of the Phanerozoic and definitively more similar to some of them than previously believed.

4.8 Summary

The Silurian global events are a new frontier of Earth science research. Three events are of particular interest because of their impact on the marine environment – the Early Silurian Ireviken Event, the Middle Silurian Mulde Event, and the Late Silurian Lau Event. These three events caused considerable extinctions and ecosystem changes in the marine environment in deep basins as well as in shallow inshore environments. Their close relationship with substantial anomalies in the global carbon cycle and changing oceanography shows that they were intimately associated with changes in the global climate, potentially a series of glaciations that today not are supported by glacial evidence such as tillites. The oceanic models that have been developed so far all suggest that the Silurian global events took place at the tipping point between stable climatic situations. The ultimate causes for the change of climate are however debated. As is evident from the paleogeographic maps presented herein, however, a majority of the studies on Silurian global events have been conducted on the Laurussian continent and from low latitudes. More studies from other continental blocks and paleolatitudes are certainly needed in order to advance the knowledge of these events. With regard to the anomalies in the global carbon cycle and the ecosystem changes in low-latitude shelves, the Silurian global events much resemble some of the 'major five' mass extinction events of the Phanerozoic.

4.9 Acknowledgement

I am grateful for the permission to use re-drawn figures from colleagues to this publication and for inspiring discussions at many occasions. Many thanks to Axel Munnecke, Oliver Lehnert, Markes E. Johnson, Mats E. Eriksson, Matthew Saltzman, Elzbieta Porębska, Anna Kozłowska, and, not least, Hanna Calner. The author's research on the Silurian global events is funded by the Swedish Research Council (VR) and Crafoord.

References

Aldridge RJ, Jeppsson L, Dorning KJ (1993) Early Silurian oceanic episodes and events. Journal of the Geological Society, London 150:501-513

Alvarez LW, Alvarez W, Asaro F, Michel HV (1980) Extraterrestrial cause for the Cretaceous-Tertiary extinction. Science 208:1095-1108

Andrew AS, Hamilton PJ, Mawson R, Talent JA, Whitford DJ (1994) Isotopic correlation tools in the Middle Palaeozoic and their relation to extinction event. APEA Journal 34:268-277

Azmy K, Veizer J, Bassett MG, Copper P (1998) Oxygen and carbon isotopic composition of Silurian brachiopods: implications for coeval seawater and glaciations. Geological Society of America Bulletin 110:1499-1512

Baarli BG, Johnson ME, Antoshkina AI (2003) Silurian stratigraphy and palaeogeography of Baltica. In Landing E, Johnson ME (eds) Silurian Lands and Seas: Paleogeography outside of Laurentia. New York State Museum Bulletin 493:3-34

Barnes C, Hallam A, Kaljo D, Kauffman EG, Walliser OH (1995) Global event stratigraphy. In Walliser OH (ed) Global events and event stratigraphy in the Phanerozoic. Springer-Verlag 319-333

Bergman CF (1989) Silurian paulinitid polychaetes from Gotland. Fossils and Strata 25:1-128

Bickert T, Pätzold J, Samtleben C, Munnecke A (1997) Palaeoenvironmental changes in the Silurian indicated by stable isotopes in brachiopod shells from Gotland, Sweden. Geochimica et Cosmochimica Acta 61:2717-2730

Brand U, Azmy K, Veizer J (2006) Evaluation of the Salinic I tectonic, Cancañiari glacial and Ireviken biotic events: biochemostratigraphy of the Lower Silurian succession in the Niagara Gorge area, Canada and USA. Palaeogeography, Palaeoclimatology, Palaeoecology 241:192-213

Brenchley PJ, Harper DAT (1998) Palaeoecology: ecosystems, environments and evolution. 402 pp. Chapman & Hall. Oxford

Calner M (1999) Stratigraphy, facies development, and depositional dynamics of the Late Wenlock Fröjel Formation, Gotland, Sweden. GFF 121:13-24

Calner M (2002) A lowstand epikarstic intertidal flat from the middle Silurian of Gotland, Sweden. Sedimentary Geology 148:389-403

Calner M (2005a) Silurian carbonate platforms and extinction events – ecosystem changes exemplified from the Silurian of Gotland. Facies 51:603-610

Calner M (2005b) A Late Silurian extinction event and anachronistic period. Geology 33:305-308

Calner M (2005c) A Late Silurian extinction event and anachronistic period: Comment and Reply. Geology Online forum e92

Calner M, Säll E (1999) Transgressive oolites onlapping a Silurian rocky shoreline unconformity, Gotland, Sweden. GFF 121:91-100

Calner M, Jeppsson L (2003) Carbonate platform evolution and conodont stratigraphy during the middle Silurian Mulde Event, Gotland, Sweden. Geological Magazine 140:173-203

Calner M, Eriksson ME (2006a) Silurian research at the crossroads. GFF 128:73-74

Calner M, Eriksson MJ (2006b) Evidence for rapid environmental changes in low latitudes during the Late Silurian Lau Event: the Burgen-1 drillcore, Gotland, Sweden. Geological Magazine 143:15-24

Calner M, Kozłowska A, Masiak M, Schmitz B (2006a) A shoreline to deep basin correlation chart for the middle Silurian coupled extinction-stable isotopic event. GFF 128:79-84

Calner M, Ahlberg P, Axheimer N, Gustavsson L (2006b) The first record of *Odontopleura ovata* (Trilobita) from Scandinavia: part of a middle Silurian intercontinental shelly benthos mass occurrence. GFF 128:37-41

Caputo MV (1998) Ordovician-Silurian glaciations and global sea-level changes. In Landing E, Johnson ME (eds) Silurian cycles: Linkages of dynamic stratigraphy with atmospheric, oceanic, and tectonic changes. New York State Museum Bulletin 491:15-25

Cocks LRM, Torsvik TH (2002) Earth geography from 500 to 400 million years ago: a faunal and palaeomagnetic review. Journal of the Geological Society, London 159:631-644

Copper P (2002) Silurian and Devonian reefs: 80 million years of global greenhouse between two ice ages. Phanerozoic reef patterns. SEPM Special Publication 72:181-238

Corfield RM, Siveter DJ, Cartlidge JE, McKerrow WS (1992) Carbon isotope excursion near the Wenlock-Ludlow (Silurian) boundary in the Anglo-Welsh area. Geology 20:371-374

Cramer BD, Saltzman MR (2005) Sequestration of 12C in the deep ocean during the early Wenlock (Silurian) positive carbon isotope excursion. Palaeogeography, Palaeoclimatology, Palaeoecology 219:333-349

Cramer BD, Saltzman MR (2007) Fluctuations in epeiric sea carbonate production during Silurian positive carbon isotope excursions: A review of proposed paleoceanographic models. Palaeogeography, Palaeoclimatology, Palaeoecology 245:37-45

Cramer BD, Kleffner MA, Saltzman MR (2006) The Late Wenlock Mulde positive carbon isotope ($\delta^{13}C_{carb}$) excursion in North America. GFF 128:85-90

Díaz-Martínez E, Grahn Y (2007) Early Silurian glaciation along the western margin of Gondwana (Peru, Bolivia and northern Argentina): Palaeogeographic and geodynamic setting. Palaeogeography, Palaeoclimatology, Palaeoecology 245:62-81

Droser ML, Bottjer DJ, Sheehan PM (1997) Evaluating the ecological architecture of major events in the Phanerozoic history of marine invertebrate life. Geology 25:167-170

Droser ML, Bottjer DJ, Sheehan PM, McGhee Jr GR (2000) Decoupling of taxonomic and ecologic severity of Phanerozoic marine mass extinctions. Geology 28:675-678

Eriksson ME (2006) The Silurian Ireviken event and vagile benthic faunal turnovers (Polychaeta; Eunicidia) on Gotland, Sweden. GFF 128:91-95

Eriksson ME, Bergman CF, Jeppsson L (2004) Silurian scolecodonts. Review of Palaeobotany and Palynology 131:269-300

Eriksson MJ, Calner M (in press) A sequence stratigraphical model for the late Ludfordian (Silurian) of Gotland, Sweden – implications for timing between changes in sea-level, palaeoecology, and the global carbon cycle. Facies

Gelsthorpe DN (2004) Microplankton changes through the early Silurian Ireviken extinction event on Gotland, Sweden. Review of Palaeobotany and Palynology 130:89-103

Gradstein FM, Ogg JG, Smith AG (2004) A geologic time scale 2004. Cambridge University Press 589 pp

Grahn Y, Caputo MV (1992) Early Silurian glaciations in Brazil. Palaeogeography, Palaeoclimatology, Palaeoecology 99:9-15

Groves JR, Calner M (2004) Lower Triassic oolites in Tethys: a sedimentologic response to the end-Permian mass extinction. Geological Society of America Abstracts with Programs 36:336

Gustavsson L, Calner M, Jeppsson L (2005) Brachiopod biodiversity changes during the Late Silurian Lau Event, Gotland, Sweden. In: Harper DAT, Long SL, McCorry M (eds) The Fifth International Brachiopod Congress, Copenhagen 2005. Abstracts, p. 40

Hagadorn JW, Bottjer DJ (1997) Wrinkle structures: Microbially mediated sedimentary structures common in subtidal siliciclastic settings at the Proterozoic-Phanerozoic transition. Geology 25:1047-1050

Hagadorn JW, Bottjer DJ (1999) Restriction of a Late Neoproterozoic biotope: Suspect microbial structures and trace fossils at the Vendian-Cambrian transition. Palaios 14:73-85

Hallam A, Wignall PB (1999) Mass extinctions and sea-level changes. Earth-Science Reviews 48:217-250
Hede JE (1921) Gottlands silurstratigrafi. Sveriges Geologiska Undersökning C305:1-100
Hints O, Killing M, Männik P, Nestor V (2006) Frequency patterns of chitinozoans, scolecodonts, and conodonts in the upper Llandovery and lower Wenlock of the Paatsalu core, western Estonia. Proceedings of the Estonian Academy of Sciences 55:128-155
Jaeger H (1991) New standard graptolite zonal sequence after the "big crisis" at the Wenlockian/Ludlovian boundary (Silurian). Neues Jahrbuch für Geologie und Paläontologie, Abhandlungen 182:303-354
Jeppsson L (1990) An oceanic model for lithological and faunal changes tested on the Silurian record. Journal of the Geological Society, London 147:663-674.
Jeppsson L (1997a) The anatomy of the mid-Early Silurian Ireviken Event. In Brett C, Baird GC (eds) Paleontological Events: Stratigraphic, Ecological, and Evolutionary Implications, 451-492. Columbia University Press, New York, N.Y
Jeppsson L (1997b) A new latest Telychian, Sheinwoodian and Early Homerian (Early Silurian) Standard Conodont Zonation. Transaction of the Royal Society of Edinburgh Earth Sciences 88:91-114
Jeppsson L (1998) Silurian oceanic events: Summary of general characteristics. In Landing E, Johnson ME (eds) Silurian cycles: Linkages of dynamic stratigraphy with atmospheric, oceanic, and tectonic changes. New York State Museum Bulletin 491: 239–257
Jeppsson L (2005) Conodont-based revisions of the Late Ludfordian on Gotland, Sweden. GFF 127:273-282
Jeppsson L, Aldridge RJ (2000) Ludlow (late Silurian) oceanic episodes and events. Journal of the Geological Society, London 157:1137-1148
Jeppsson L, Calner M (2003) The Silurian Mulde Event and a scenario for secundo-secundo events. Transaction of the Royal Society of Edinburgh Earth Sciences 93:135-154
Jeppsson L, Aldridge RJ, Dorning KJ (1995) Wenlock (Silurian) oceanic episodes and events. Journal of the Geological Society, London 152:487–498
Jeppsson L, Eriksson ME, Calner M (2006) The Silurian high-resolution stratigraphy of Gotland – a summary. GFF 128:109-114
Jeppsson L, Talent JA, Mawson R, Simpson AJ, Andrew A, Calner M, Whitford D, Trotter JA, Sandström O, Caldon HJ (2007) High-resolution Late Silurian correlations between Gotland, Sweden, and the Broken River region, NE Australia: lithologies, conodonts and isotopes. Palaeogeography, Palaeoclimatology, Palaeoecology 245:115-137
Johnson ME (2006) Relationship of Silurian sea-level fluctuations to oceanic episodes and events. GFF 128:115-121
Kaljo D, Boucot AJ, Corfield RM, Le Herisse A, Koren TN, Kříž J, Männik P, Märss T, Nestor V, Shaver RH, Siveter DJ, Viira V (1995) Silurian bio-events. In Walliser OH (ed) Global events and event stratigraphy in the Phanerozoic. Springer-Verlag 173-224

Kaljo D, Kiipli T, Martma T (1997) Correlation of carbon isotope event markers through the Wenlock-Pridoli sequence at Ohesaare (Estonia) and Priekule (Latvia). Palaeogeography, Palaeoclimatology, Palaeoecology 132:211–223

Kaljo D, Martma T, Männik P, Viira V (2003) Implications of Gondwana glaciations in the Baltic late Ordovician and Silurian and a carbon isotopic test of environmental cyclicity. Bull. Soc. Géol. Fr. 174:59-66

Kaljo D, Martma T, Neuman BEE, Rønning K (2004) Carbon isotope dating of several uppermost Ordovician and Silurian sections in the Oslo region, Norway. Wogogob meeting abstract 51-52

Kaljo D, Martma T (2006) Application of carbon isotope stratigraphy to dating Baltic Silurian rocks. GFF 128:161-168

Koren' TN (1991) The *lundgreni* extinction event in central Asia and its bearing on graptolite biochronology within the Homerian. Proceedings of the Estonian Academy of Sciences, Geology 40:74-8

Koren' TN (1993) Main event levels in the evolution of the Ludlow graptolites. Geological Correlation 1:44-52

Kozłowska-Dawidziuk A, Lenz AC, Storch P (2001) Upper Wenlock and Lower Ludlow (Silurian), post-extinction graptolites, Vseradice section, Barrandian Area, Czech Republic. Journal of Paleontology 75:147-164

Kozłowski W (2003) Age, sedimentary environment and palaeogeographical position of the Late Silurian oolitic beds in the Holy Cross Mountains (Central Poland). Acta Geologica Polonica 53:341-357

Laufeld S (1974) Silurian chitinozoa from Gotland. Fossils and Strata 5:1-130

Lehnert O, Fryda J, Buggisch W, Manda S (2003) A first report of the Ludlow Lau event from the Prague Basin (Barrandian, Czech Republic). In Insugeo, G Ortega, GF Aceñolaza (eds) Serie Correlación Geológica 18:139-144

Lehnert O, Eriksson MJ, Calner M, Joachimski M, Buggisch W (2007a) Concurrent sedimentary and isotopic indications for global climatic cooling in the Late Silurian. Acta Palaeontologica Sinica 46:249-255

Lehnert O, Fryda J, Buggisch W, Munnecke A, Nützel, Kříž J, Manda S (2007b) $\delta^{13}C$ records across the late Silurian Lau event: New data from middle palaeo-latitudes of northern peri-Gondwana (Prague Basin, Czech Republic) Palaeogeography, Palaeoclimatology, Palaeoecology 245:227-244

Lenz AC, Kozłowska-Dawidziuk A (2001) Upper Wenlock (Silurian) graptolites of the Arctic Canada: pre-extinction, *lundgreni* Biozone fauna. Palaeontographica Canadiana 20:1-61

Lenz AC, Kozłowska A (2006) Graptolites from the *lundgreni* Biozone (lower Homerian, Silurian), Arctic Islands, Canada: new species and supplementary material. Journal of Paleontology 80:616-637

Lenz AC, Noble PJ, Masiak M, Poulson SR, Kozłowska A (2006) The *lundgreni* Extinction Event: integration of paleontological and geochemical data from Arctic Canada. GFF 128:153-158

Loydell DK, Kaljo D, Männik P (1998) Integrated biostratigraphy of the lower Silurian of the Ohesaare core, Saaremaa, Estonia. Geological Magazine 135: 769-783

Loydell DK, Männik P, Nestor V (2003) Integrated biostratigraphy of the lower Silurian of the Aizpute-41 core, Latvia. Geological Magazine 140: 205-229.

Manda S, Kříž J (2006) Environmental and biotic changes in subtropical isolated carbonate platforms during the Late Silurian Kozlowskii Event, Prague Basin. GFF 128:161-168

Märss T (1992) Vertebrate history in the Late Silurian. Proceedings of the Estonian Academy of Sciences, Geology 41:205-214

Martinsson A (1967) The succession and correlation of ostracode faunas in the Silurian of Gotland. Geologiska Föreningens i Stockholm Förhandlingar 89:350-386

Melchin JM, Koren TN, Storch P (1998) Global diversity and survivorship patterns of Silurian graptoloids. In Landing E, Johnson ME (eds) Silurian cycles: Linkages of dynamic stratigraphy with atmospheric, oceanic, and tectonic changes. New York State Museum Bulletin 491:165-181

Mullins GL, Aldridge RJ (2004) Chitinozoan biostratigraphy of the basal Wenlock Series (Silurian) Global Stratotype Section and Point. Palaeontology 47:745-773

Munnecke A, Samtleben C, Bickert T (2003) The Ireviken Event in the lower Silurian of Gotland, Sweden – relation to similar Palaeozoic and Proterozoic events. Palaeogeography, Palaeoclimatology, Palaeoecology 195:99-124

Nestor V, Einasto R, Loydell DK (2002) Chitinozoan biostratigraphy and lithological characteristics of the Lower and Upper Visby boundary beds in the Ireviken 3 section, Northwest Gotland. Proceedings of the Estonian Academy of Sciences, Geology 51:215-226

Noble PJ, Zimmerman MK, Holmden C, Lenz AC (2005) Early Silurian (Wenlockian) $\delta^{13}C$ profiles from the Cape Phillips Formation, Arctic Canada and their relation to biotic events. Canadian Journal of Earth Sciences 42:1419-1430

Noble P, Lenz A, Holmden C, Masiak M, Poulson SR, Zimmerman MK, Kozłowska A (2006) Integrated paleontologic, sedimentologic, and stable isotopic data across the Ireviken and *lundgreni* extinction events in the Cape Phillips Formation, Nunavut, Canada. Geological Society of America Abstracts with Programs 38:552

Pittau P, Cotza F, Cristini S, Del Rio M, Loi M (2006) Palaeontologic and biogeochemical characterization of the *Cyrtograptus lundgreni* event in the black shales of eastern Mid-Sardinia, Italy. Lethaia 39:111-127

Porębska E, Kozłowska-Dawidziuk A, Masiak M (2004) The *lundgreni* event in the Silurian of the East European Platform, Poland. Palaeogeography, Palaeoclimatology, Palaeoecology 213:271-294

Pruss S, Fraiser M, Bottjer DJ (2004) Proliferation of Early Triassic wrinkle structures: Implications for environmental stress following the end-Permian mass extinction. Geology 32:461-464

Puura V, Ainsaar L, Eriksson M, Flodén T, Kirsimäe K, Kleesment A, Konsa M, Suuroja K (2000) Known and unknown meteorite impact events, as recorded in the epicontinental Lower Palaeozoic sedimentary sequence of the Baltic region. Catastrophic events conference, Vienna 2000, Abstract 3083

Ramsköld L (1985) Studies on Silurian trilobites from Gotland, Sweden. Department of Geology, University of Stockholm, and Department of Palaeozoology, Swedish Museum of Natural History, Stockholm, 24 pp

Raup DM, Sepkoski JJ Jr (1982) Mass extinctions in the marine fossil record. Science 215:1501-1503

Rickards RB, Packham GH, Wright AJ, Williamson PL (1995) Wenlock and Ludlow graptolite faunas and biostratigraphy of the Quarry Creek district, New South Wales. Association of Australasian Palaeontologists, Memoir 17:1-68

Riding R, Liang L (2005) Geobiology of microbial carbonates: Metazoan and seawater saturation state influences on secular trends during the Phanerozoic. Palaeogeography, Palaeoclimatology, Palaeoecology 219:101-115

Saltzman MR (2001) Silurian $\delta 13C$ stratigraphy: A view from North America. Geology 29:671-674

Saltzman MR (2005) Phosphorous, nitrogen, and the redox evolution of the Paleozoic oceans. Geological Society of America Bulletin 33:573-576

Samtleben C, Munnecke A, Bickert T, Pätzold J (1996) The Silurian of Gotland (Sweden): Facies interpretation based on stable isotopes in brachiopod shells. Geologische Rundschau 85:278-292

Samtleben C., Munnecke A, Bickert T (2000) Development of facies and C/O-isotopes in transects through the Ludlow of Gotland: Evidence for global and local influences on a shallow-marine environment. Facies 43:1-38

Schönlaub HP (1986) Significant geological events in the Paleozoic record of the Southern Alps (Austrian part). In Walliser OH (ed) Global bio-events 163-167. Springer-Verlag

Schubert JK, Bottjer DJ (1992) Early Triassic stromatolites as post-mass extinction disaster forms. Geology 20:883-886

Sepkoski JJ Jr (1982) Flat-pebble conglomerates, storm deposits, and the Cambrian bottom fauna. In Einsele G, Seilacher A (eds) Cyclic and event stratification, 371-385. Spriner-Veerlag, Berlin

Sheehan PM, Harris MT (2004) Microbialite resurgence after the Late Ordovician extinction. Nature 430:75-77

Storch P (1995) Biotic crisis and post-crisis recoveries recorded by Silurian planktonic graptolite faunas of the Barrandian Area (Czech Republic). Geolines 3:59-70

Talent JA, Mawson R, Andrew AS, Hamilton PJ, Whitford DJ (1993) Middle Palaeozoic events: Faunal and isotopic data. Palaeogeography, Palaeoclimatology, Palaeoecology 104:139-152

Urbanek A (1970) Neocucullograptinae n. subfam. (Graptolithina) – Their evolutionary and stratigraphic bearing. Acta Palaeontologica Polonica 15:1-388

Urbanek A (1993) Biotic crises in the history of Upper Silurian graptoloids: A palaeobiological model. Historical Biology 7:29-50

Valentine JL, Brock GA, Molloy PD (2003) Linguliformean brachiopod faunal turnover across the Ireviken Event (Silurian) at Boree Creek, central-western New South Wales, Australia. Cour. Forsch.-Inst. Senckenberg, 242:301-327

Walliser OH (1964) Conodonten des Silurs. Abhandlungen des Hessischen Landesamtes für Bodenforschung 41:1-106

Wenzel B (1997) Isotopenstratigraphische Untersuchungen an silurischen Abfolgen und deren paläozeanographische Interpretation. Erlanger Geologische Abhandlungen 129:1-117

Whalen MT, Day J, Eberli GP, Homewood PW (2002) Microbial carbonates as indicators of environmental change and biotic crises in carbonate systems: examples from the Late Devonian, Alberta basin, Canada. Palaeogeography, Palaeoclimatology, Palaeoecology 181:127-141

Wigforss-Lange J (1999) Carbon isotope $\delta^{13}C$ enrichment in Upper Silurian (Withcliffian) marine calcareous rocks in Scania, Sweden. GFF 121:273-279

5 Late Devonian mass extinction

Ashraf M. T. Elewa

Geology Department, Faculty of Science, Minia University, Minia 61519, Egypt, aelewa@link.net

The Devonian mass extinction occurred during the latter part of the Devonian at the Frasnian-Famennian boundary. The crisis primarily affected the marine community, having little impact on the terrestrial flora. This same extinction pattern has been recognized in most mass extinctions throughout Earth history. The most important group to be affected by this extinction event was the major reef-builders including the stromatoporoids, and the rugose, and tabulate corals. Among other marine invertebrates, seventy percent of the taxa did not survive into the Carboniferous (see the Wikipedia, and McGhee 1996).

The Hooper Museum (the website of the Hannover Park, Hooper Virtual Paleontological Museum) ascribed this event either to glaciation or to meteorite impact. They cited, for example, that warm water marine species were the most severely affected in this extinction event. This evidence has lead many paleontologists to attribute the Devonian extinction to an episode of global cooling, similar to the event that is thought to cause the late Ordovician mass extinction. According to this theory, the extinction of the Devonian was triggered by another glaciation event on Gondwana, as evidenced by glacial deposits of this age in northern Brazil. On the other hand, the museum clued that meteorite impacts at the Frasnian-Famennian boundary have also been suggested as possible agents for the Devonian mass extinction.

An important note has been introduced by George R. McGhee Jr (1996), who has detected, among the survivors, some trends that lead to his conclusion that survivors generally represent more primitive or ancestral morphologies. In other words, the conservative generalists are more likely to survive an ecological crisis than species that have evolved as specialists.

Prothero (1998) believes in a severe cooling event for the Late Devonian mass extinction.

Joachimski and Buggisch (2000) suggested that climatic cooling in conjunction with significant oceanographic changes could represent a powerful scenario to account for the Late Devonian mass extinction event.

However, the Devonian Times website pointed out that the timing and duration of the Late Devonian mass extinction(s) are subject to considerable debate and a variety of interpretations.

Generally, several causes have been suggested for the Devonian mass extinction event, including asteroid impacts, global anoxia, plate tectonics, sea level changes, and climatic change. Yet, one of the most attractive of these is the "Devonian Plant Hypothesis" of Algeo et al. (1995). These authors judged the extension of terrestrial plants as the decisive cause for mass extinction in the tropical oceans.

References

Algeo TJ, Berner RA, Maynard JP, Scheckler SE (1995) Late Devonian oceanic anoxic events and biotic crises: "Rooted" in the evolution of vascular land plants? GSA Today 5 (45): 64-66

Joachimski MM, Buggisch W (2000) The Late Devonian mass extinction - Impact or Earth bound event? Catastrophic Events and Mass Extinctions: Impacts and Beyond (Abstracts), Lunar and planetary Institute Meeting, Houston, USA

McGhee GR Jr (1996) The Late Devonian Mass Extinction: The Frasnian/Famennian Crisis. New York: Columbia University Press, USA

Prothero DR (1998) Bringing fossils to life: An introduction to paleobiology. WCB/McGrow-Hill, USA

6 Late Permian mass extinction

Ashraf M. T. Elewa

Geology Department, Faculty of Science, Minia University, Minia 61519, Egypt, aelewa@link.net

The Permian-Triassic (P-T or PT) extinction event, sometimes informally called the Great Dying, was an extinction event that occurred approximately 251 million years ago (mya), forming the boundary between the Permian and Triassic geologic periods. It was the Earth's most severe extinction event, with about 96 percent of all marine species and 70 percent of terrestrial vertebrate species becoming extinct. Many theories have been presented for the cause of the extinction, including plate tectonics, an impact event, a supernova, extreme volcanism, the release of frozen methane hydrate from the ocean beds to cause a greenhouse effect, or some combination of factors. Recently, a group of scientists claimed they have discovered a 300-mile-wide (480 km) crater in the Wilkes Land region of East Antarctica, which they believe may be linked with the extinction (Wikipedia, the free encyclopedia).

The Hooper Museum (the website of the Hannover Park, Hooper Virtual Paleontological Museum) stated that the Permian mass extinction occurred about 248 million years ago and was the greatest mass extinction ever recorded in earth history; even larger than the previously discussed Ordovician and Devonian crises and the better known End Cretaceous extinction that felled the dinosaurs. Ninety to ninety-five percent of marine species were eliminated as a result of this Permian event. The primary marine and terrestrial victims included the fusulinid foraminifera, trilobites, rugose and tabulate corals, blastoids, acanthodians, placoderms, and pelycosaurs, which did not survive beyond the Permian boundary. Other groups that were substantially reduced included the bryozoans, brachiopods, ammonoids, sharks, bony fish, crinoids, eurypterids, ostracodes, and echinoderms.

The museum assigned this mass extinction to one of the following causes:
1. Global widespread cooling and/or worldwide lowering of sea level
2. The formation of Pangaea
3. Glaciation
4. Volcanic eruptions

Prothero (1998) declared that although we can rule out extraterrestrial impacts, reduced marine shelf habitat, and global cooling as causes, a great number of other environmental stresses were clearly operating at the end of the Permian.

The National Geographic Magazine summarized the problem in the words said by Doug Erwin of the Smithsonian, who said: "The truth is sometimes untidy." The Permian extinction reminds him of Agatha Christie's Murder on the Orient Express, in which a corpse with 12 knife wounds is discovered on a train. Twelve different killers conspired to slay the victim. Erwin suspects there may have been multiple killers at the end of the Permian. May be everything (eruptions, an impact, and anoxia) gone wrong at once.

Reference

Prothero DR (1998) Bringing fossils to life: An introduction to paleobiology. WCB/McGrow-Hill, USA, 560 pp

7 Late Triassic mass extinction

Ashraf M. T. Elewa

Geology Department, Faculty of Science, Minia University, Minia 61519, Egypt, aelewa@link.net

The Triassic-Jurassic extinction event occurred 200 million years ago and is one of the major extinction events of the Phanerozoic eon, profoundly affecting life on land and in the oceans. 20% of all marine families and all large Crurotarsi (non-dinosaurian archosaurs), some remaining therapsids, and many of the large amphibians were wiped out. At least half of the species now known to have been living on Earth at that time went extinct (see the Wikipedia).

The Wikipedia revealed that recently some evidence has been retrieved from near the Triassic-Jurassic boundary suggesting that there was a rise in atmospheric CO_2 and some researchers have suggested that the cause of this rise, and of the mass extinction itself, could have been a combination of volcanic CO_2 outgassing and catastrophic dissociation of gas hydrate.

The University of Washington, in May 10, 2001, declared that a collapse of simple life forms(one-celled organisms called protests) linked to mass extinction 200 million years ago (e.g. the Late Triassic mass extinction event).

The SPACE.com website, in May 11, 2001, stated that the fifth worst mass extinction linked to asteroid impact.

The Late Triassic website of Bristol University introduced the following theories on this mass extinction event:
1. Fluctuating sea level change
2. The Earth got hit by a large rock
3. Death by volcano
4. Climate change

The severe effect on Triassic reefs, especially in the Tethys, suggests that cooling was a significant factor (Prothero 1998). Prothero added that the abundance of black shales and the geochemical anomalies suggest that major oceanic changes were important.

On the other hand, Olsen et al. (1987) clued the possibility that impacts may played a role on this event.

Some scientists, however, believe that there were no mass extinctions in the Late Triassic!!

References

Olsen PE, Shubin NH, Anders MH (1987) New Early Jurassic tetrapod assemblages constrain Triassic-Jurassic tetrapod extinction event. Science 237: 1025-1029

Prothero DR (1998) Bringing fossils to life: An introduction to paleobiology. WCB/McGrow-Hill, USA, 560 pp

8 Reexamination of the end-Triassic mass extinction

Spencer G. Lucas[1] and Lawrence H. Tanner[2]

[1]New Mexico Museum of Natural History, 1801 Mountain Road NW, Albuquerque, New Mexico 87104-1375 US, spencer.lucas@state.nm.us; [2]Department of Biological Sciences, Le Moyne College, 1418 Salt Springs Road, Syracuse, NY 13214 USA, tannerlh@lemoyne.ed

8.1 Introduction

The biodiversity crisis at the end of the Triassic (Triassic-Jurassic boundary: TJB) has long been identified as one of the "big five" mass extinctions of the Phanerozoic. Attribution of this level of suddenness and severity to the TJB extinction began with Sepkoski (1982), who, based on a global compilation of families of marine invertebrates, designated the TJB extinction as one of four mass extinctions events of intermediate magnitude (end-Cretaceous, end-Triassic, Late Devonian, Late Ordovician), less severe than the largest Phanerozoic extinction, which was at the end of the Permian. This identification of a severe and sudden biotic decline at the TJB remained unquestioned until recently (Hallam 2002; Tanner et al. 2004; Lucas and Tanner 2004).

Here, we review the magnitude and timing of the extinctions that took place during the Late Triassic. Our thesis is that the Late Triassic was an interval of elevated extinction rates and low origination rates that manifested themselves in a series of discrete extinctions during Carnian, Norian and Rhaetian time. Significantly, no reliable data exist to document global Late Triassic mass extinction(s) of many significant biotic groups, including foraminiferans, ostracods, gastropods, fishes and marine reptiles (Hallam 2002; Tanner et al. 2004). Therefore, we focus our discussion on those groups that have been perceived by some as part of a TJB mass extinction, namely radiolarians, marine bivalves, ammonoids, reef-building organisms, conodonts, land plants and terrestrial tetrapods (amphibians and reptiles).

8.2 Late Triassic Timescale

We use the Late Triassic timescale compiled by Ogg (2004), although we attribute different numerical ages to the stage boundaries based on new data, such as Furin et al. (2006) (Fig. 1). The bases of the three Late Triassic stages (Carnian, Norian and Rhaetian) have not been formally defined by GSSPs (global stratotype sections and points), so we use traditional definitions for the bases of the Carnian and Norian (see Ogg 2004). However, we follow the proposal of Krystyn et al. (2007) and recognize a Rhaetian that encompasses two ammonoid zones. This means the Rhaetian includes the latter part of Sevatian 2; hence our usage is of the so-called "long" Rhaetian (also see Gallett et al. 2007).

The long-accepted definition of the TJB (base of Jurassic System = base of Hettangian Stage) has been the lowest occurrence of the smooth-shelled psiloceratid ammonoid *Psiloceras planorbis* in southern England (e.g. Lloyd 1964; Maubeuge 1964; Cope et al. 1980; Warrington et al. 1994; Ogg 2004; Gradstein et al. 2004). However, current work under the guidelines of the IUGS International Commission on Stratigraphy seeks ratification of a GSSP for that boundary (Warrington 2005). Criteria for possible boundary definition (Fig. 2) include a negative carbon isotope excursion, a bivalve bio-event, an evolutionary turnover in radiolarians, continued use of the lowest occurrence of *P. planorbis*, or the lowest occurrence of ammonoids of the *P. tilmanni* group (Lucas et al. 2005, 2006, 2007a). We favor and employ the definition of the base of the Jurassic at the lowest occurrence of ammonoids of the *Psiloceras tilmanni* group (Guex et al. 2004; Lucas et al. 2007a).

Correlation of nonmarine and marine biochronology in the Late Triassic remains imprecise. Here, we use the correlations advocated by Lucas et al. (2007b) and Lucas and Tanner (2007b).

8.3 Some methodological issues

Two methods have been used to analyze biodiversity changes at the TJB: (1) the compilation of global diversity from the published literature; and (2) the study of diversity changes based on the actual distribution of fossils in specific stratigraphic sections. These two methods are not totally disjunct, because the global compilations are based on the actual stratigraphic distributions of the fossils in all sections. However, the global compilations contain a serious flaw--their stratigraphic (temporal) imprecision (Teichert, 1988), which Lucas (1994) termed the compiled correlation effect. We believe that this imprecision is largely responsible for the concept of a single, TJB mass extinction.

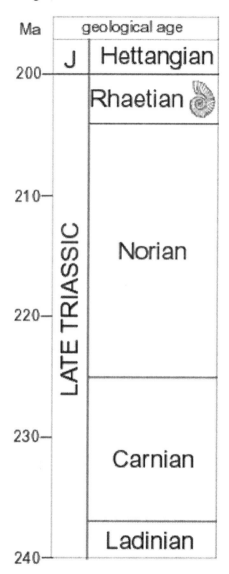

Fig. 1. The Late Triassic timescale

The compiled correlation effect refers to the fact that the temporal ranges of taxa in literature compilations are only as precise as the correlations, or relative ages, of the compiled taxa. Because most published correlations are at the stage/age level, the temporal resolution of extinction events within these stages/ages cannot be resolved. The result is the artificial concentration of extinctions at stage/age boundaries. A complex extinction of significant temporal duration during a stage/age can be made to appear as a mass extinction at the end of the stage/age (Fig. 3).

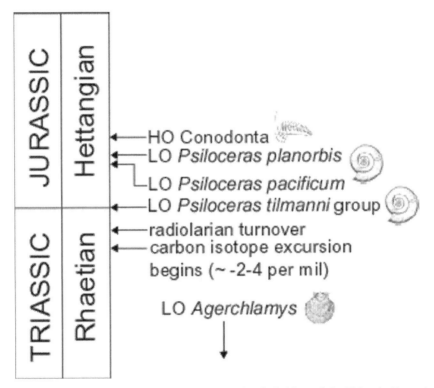

Fig. 2. Succession of potential marker events for definition of the Triassic-Jurassic boundary (after Lucas et al., 2005). HO = highest occurrence, LO = lowest occurrence

Much of the literature on the TJB extinction has failed to consider the compiled correlation effect. Thus, for example, the supposedly profound extinction of ammonoids at the end of the Rhaetian reflects a lack of detailed stratigraphic analysis; literature compilations assume that any ammonoid taxon found in Rhaetian strata has a stratigraphic range throughout the entire Rhaetian (e.g. House 1989). This gives the appearance of a dramatic ammonoid extinction at the end of the Rhaetian, when in fact, ammonoid taxa were experiencing extinction throughout the Rhaetian (Fig. 3). A very recent analysis heavily influenced by the compiled correlation effect is that of Kiessling et al. (2007), who used the Paleobiology Database to evaluate the TJB extinction. They thus compared one Rhaetian diversity point to one Hettangian point and concluded there was a "true mass extinction" (Kiessling et al. 2007: 220) at the TJB.

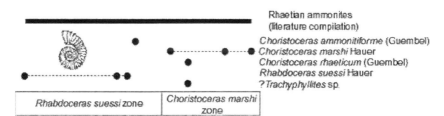

Fig. 3. The actual ranges of Rhaetian ammonites in the Weissloferbach section (Austria) of the Kössen Beds (after Mostler et al., 1978) show a low diversity Rhaetian ammonoid assemblage with only one taxon (*Choristoceras marshi*) present at the top of the Rhaetian section. In contrast, the low stratigraphic resolution characteristic of literature compilations indicates all ammonoid ranges simply truncated at the top of the Rhaetian, a typical example of the compiled correlation effect

We also note that the Signor-Lipps effect (Signor and Lipps 1982) has been used by some to discount the reliability of actual stratigraphic ranges. This theory suggests that some actual stratigraphic ranges in the fossil record are artificially truncated by incomplete sampling, and statistical methods exist to "complete" these supposedly truncated stratigraphic ranges. However, we regard use of these methods as little more than assumptions that invent data, and prefer to rely on the actual stratigraphic ranges of fossils in well-studied sections (Fig. 4). Indeed, a caution to those who would use statistical methods of stratigraphic range estimation to analyze extinctions is provided by the statistical analyses of Ward et al. (2005) and Marshall (2005), which used the same dataset of taxon ranges relevant to the end-Permian extinctions to support different conclusions based on different assumptions.

Another problem in analyzing the TJB extinctions stems from comparing local events in actual stratigraphic sections to broader global patterns. For example, one of the most studied marine TJB sections is at St. Audrie's Bay in southern England (e.g. Hesselbo et al. 2002, 2004; Hounslow et al. 2004). In this section, there is a major facies change that reflects a substantial marine transgression that began very close to the beginning of the Jurassic. Yet, many studies (e.g. Barras and Twitchett 2007; Schootbrugge et al. 2007) have documented biotic changes in the St Audrie's Bay section and inferred that they reflect global events, when in fact they may more simply be explained by local facies changes.

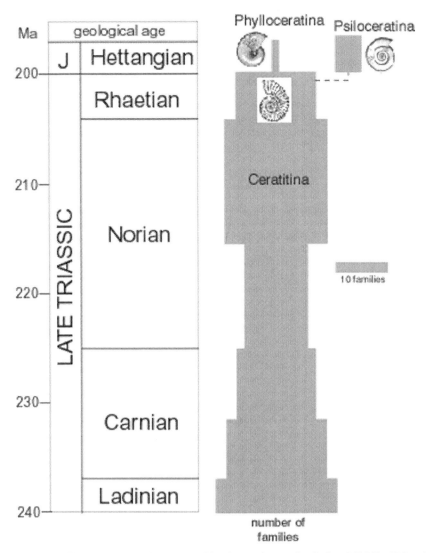

Fig. 4. Family diversity of ammonoids from the end of the Middle Triassic (Ladinian) through the earliest Jurassic (Hettangian) (based on Tozer, 1981), and simplified phylogeny of ammonoids across the TJB (based on Rakús 1993 and Guex 2006). The post-Triassic ammonoids diversified from the Psiloceratina

On the other side of this issue, global data sometimes show no extinction of a biotic group across the TJB, whereas well-analyzed local (regional) data show otherwise. Tomašových and Siblík (2007) present an example of this with their excellent documentation of major changes in the brachiopod communities across the TJB in the Northern Calcareous Alps (Austria), whereas global data suggest no substantial extinction of brachiopods across the TJB (Hallam 2002; Tanner et al. 2004). One explanation of this may be that the profound facies change across the TJB that occurs in the Northern Calcareous Alps is correlated to (underlies) the brachiopod changes, but the possibility that this well analyzed record is a more sensitive determinant of a global change needs to be considered and further evaluated.

8.4 Extinctions of taxonomic groups

8.4.1 Radiolarians

Among major faunal groups, the Radiolaria are now considered by some as optimal for defining the TJB (Carter 1994; Ward et al. 2001; Longridge et al. 2007). However, understanding the nature of the timing and severity of radiolarian extinction at the TJB has been hampered by slow identification of suitable and correlatable sections on a global scale. Blome (1986), for example, found that Tethyan and North American assemblages differed significantly at the species level, preventing direct correlation. Hence, the uppermost Triassic (Rhaetian) was characterized by the *Globotaxtorum tozeri* Zone in North America, the *Livarella densiporata* Zone in Europe, the *Canoptum triassicum* Zone in Siberia, and the *Betraccium deweveri* Zone in Japan (reviewed in Blome at al. 1995).

We argued previously that the data on the radiolarian extinction failed to demonstrate that it was a global event. Thus, at the family level, radiolarians show no serious decline at the TJB (Hart and Williams 1993), although a significant species turnover is indicated. Hori (1992), from the study of bedded cherts in central Japan, advocated a gradual end-Triassic radiolarian turnover, a conclusion shared by Vishnevskaya (1997), who demonstrated that about 40% of the latest Triassic radiolarian genera survived the TJB. Indeed, a second very large radiolarian extinction occurred later, during the Early Jurassic (early Toarcian) (Racki 2003). Furthermore, occurrences of bedded cherts show no decrease from the Late Triassic to the Early Jurassic, suggesting that there was no significant radiolarian decline (Kidder and Erwin 2001).

Nevertheless, a rapidly growing global database ably summarized by Carter (2007) indicates otherwise and supports the idea of a drastic and rapid evolutionary turnover of radiolarians across the TJB. Indeed, it has been just within the last two decades that sections with sufficiently global distribution have been studied to allow more definitive species correlation among these regions, and permit clearer interpretation of the radiolarian record across the TJB.

The best-studied and most complete radiolarian record across the TJB is in the Queen Charlotte Islands in western Canada. The Rhaetian radiolarian fauna here includes over 160 species (Carter 1993, 1994), many of which have now been identified in sections from such diverse localities as Baja California Sur (Mexico), the Philippines, China, Tibet, Russia, the southern Apennines (Italy), Turkey, and Hungary (Carter 2007). Carter (1993, 1994) established the *Proparvicingula moniliformis* Zone and the *Globolaxtorum tozeri* Zone to encompass the lower and upper Rhaetian radiolarian assemblages, respectively, in the Queen Charlotte Islands. Over half of the species present at the base of the *P. moniliformis* Zone disappear by the top of this zone, but most of the 70-plus species present at the base of the *G. tozeri* Zone continue to the system boundary.

A drastic extinction of radiolarians at the TJB was first indicated by the data in the Queen Charlotte Islands (Tipper et al. 1994; Carter 1994; Ward et al. 2001; Longridge et al. 2007). Carter (1994), for example, documented the loss of 45 radiolarian species in the top 1.5 m of the *Globolaxtorum tozeri* zone (topmost Rhaetian) on Kunga Island in the Queen Charlotte Islands. In total, it appears that 5 families, 25 genera, and most species of the *G. tozeri* Zone disappear within just a few meters of section (Carter, 1994; Longridge et al., 2007), and a similar pattern is now interpreted from Japan, where 20 genera and 130 Rhaetian species disappear across the TJB (Carter and Hori 2005).

The extinction is marked by the loss of the most architecturally complex forms of spumellarians, nassellarians, and enactiniids. The succeeding fauna is a low diversity Hettangian assemblage of morphologically conservative forms in which nassellarians are rare. Carter and Hori (2005) drew attention to how this parallels the ammonoid turnover at the TJB (complex to simple, high diversity to low diversity; see below) and argued that a short and severe environmental stress caused the radiolarian extinction across the TJB. Longridge et al. (2007) explored this point further and noted that the temporary persistence of some Rhaetian forms suggests that the extinction, while rapid, was not instantaneous. Further, they noted that the abundance of some opportunists, such as *Archaeocenosphaera laseekensis*, demonstrates rapid restoration of marine productivity. Thus, there was a significant evolutionary turnover of radiolarians at or very close to the TJB, and this appears to have been a global event tied to an abrupt and short-term decline in marine productivity.

8.4.2 Marine bivalves

The idea of a single mass extinction of marine bivalves at the end of the Triassic stems from Hallam (1981), who claimed a 92% extinction of bivalve species at the TJB. He based this estimate on combining all Norian (including Rhaetian) marine bivalve taxa into one number, thereby encompassing a stratigraphic interval with a minimum duration of 20 million years (Fig. 1). He then compared this to a single number of Hettangian marine bivalve diversity, thus providing a clear example of the compiled correlation effect.

Johnson and Simms (1989) pointed out that much better stratigraphic resolution could be achieved on the local scale; in the Kössen beds, for example, Hallam considered all of the marine bivalve taxa to range throughout the Rhaetian, even though published data (e.g. Morbey 1975) showed varied highest occurrences throughout the Rhaetian section. Furthermore, Skelton and Benton's (1993) global compilation of marine bivalve family ranges showed a TJB extinction of 5 families, with 52 families passing through the boundary unscathed, certainly suggesting that there was not a mass extinction of bivalve families.

Hallam and Wignall (1997) re-examined the marine bivalve record for the TJB in northwestern Europe and the Northern Calcareous Alps in considerable detail. They found extinction of only 4 out of 27 genera in northwest Europe and 9 of 29 genera in the Northern Calcareous Alps, again, not indicating a mass extinction. Although Hallam (2002) continued to argue for a substantial TJB marine bivalve extinction, he conceded that the data to demonstrate this are not conclusive.

Indeed, a review of the Late Triassic marine bivalve record suggests that extinctions were episodic throughout this interval, not concentrated at the TJB. A significant extinction of bivalves, including the virtual disappearance (two dwarf Rhaetian species are now known: McRoberts 2007; Krystyn et al. 2007) of the cosmopolitan and abundant pectinacean *Monotis,* is well documented for the end Norian (Dagys and Dagys 1994; Hallam and Wignall 1997). Thus, McRoberts' (2007) summary of the Late Triassic diversity dynamics of "flat clams" (halobiids and monotids) indicates they suffered their largest extinction at the Norian-Rhaetian boun-dary. The end-Norian extinction of megalodontid bivalves was noted by Allasinaz (1992), who concluded that the end-Norian marine bivalve extinction was larger than the end-Rhaetian (TJB) extinction.

Detailed studies of Late Triassic marine bivalve stratigraphic distributions (e.g. Allasinaz 1992; McRoberts 1994; McRoberts and Newton 1995; McRoberts et al. 1995) identify multiple and selective bivalve extinction events within the Norian and Rhaetian Stages and across the TJB. However, in many sections, particularly in Europe, changes in bivalve diversity and composition correlate to facies changes, and this compromises interpretation of the broader significance of these changes (Allasinaz 1992). The pattern of marine bivalve extinction during the Late Triassic is one of multiple extinction events, with a particularly significant extinction at the Norian-Rhaetian boundary, not a single mass extinction at the TJB.

8.4.3 Ammonoids

Biostratigraphic recognition (and definition) of the TJB has long been based on a clear change in the ammonoid fauna from the diverse and ornamented ceratites and their peculiar heteromorphs of the Late Triassic to the less diverse and smooth psiloceratids of the Early Jurassic (Fig. 5). This is the extinction of the Ceratitida followed by the diversification of the Ammonitida (e.g. House 1989). Most workers agree that all but one lineage of ammonites (the Phylloceratina) became extinct by the end of the Triassic, and the subsequent Jurassic diversification of ammonites evolved primarily from that lineage (Guex 1982, 1987, 2001, 2006; Rakús 1993) (Fig. 5). House (1989: 78) considered the end-Triassic ammonoid extinction "the greatest in the history of the Ammonoidea."

Indeed, there is a substantial evolutionary turnover in the ammonoids across the TJB, and Early Jurassic ammonoid assemblages are qualitatively very different from Late Triassic assemblages. The Early Jurassic encompasses a complex and rapid re-diversification of the ammonoids, from a medium-sized ancestry (e.g. Rakús 1993; Dommergues et al. 2001, 2002; Sandoval et al. 2001; Guex 2001, 2006). However, there is a strong correlation between Triassic and Jurassic ammonoid diversity and global sea-level curves that indicate ammonoid diversity crashes correspond to sea level falls (Sandoval et al. 2001). Thus, Kennedy (1977) and Signor and Lipps (1982) correlated the drop in ammonoid diversity at the end of the Triassic with a drop in sedimentary rock area, not with a mass extinction. Furthermore, Teichert (1988) listed more than 150 ammonite genera and subgenera during the Carnian, which was reduced to 90 in the Norian, and reduced again to 6 or 7 during the Rhaetian. This indicates that the most severe drops in ammonoid diversity took place during or at the end of the Carnian and Norian, not at the end of the Rhaetian (also see Tozer 1981 and Benton 1986, 1991).

The most completely studied and ammonoid-rich section in the world that crosses the TJB is in the New York Canyon area of Nevada, USA (Fig. 4). Taylor et al. (2000, 2001), Guex et al. (2002, 2003) and Lucas et al. (2007a) plotted ammonoid distribution in this section based on decades of collecting and study; of 11 Rhaetian species, 7 extend to the upper Rhaetian, and only 1 is present at the stratigraphically highest Rhaetian ammonite level. Taylor et al. (2000) presented a compelling conclusion from these data: a two-phase latest Triassic ammonoid extinction, one in the late Norian followed by a low diversity Rhaetian ammonoid fauna that became extinct by the end of the Triassic.

Another detailed study of latest Triassic ammonoid distribution is in the Austrian Kössen Beds (Ulrichs 1972; Mostler et al. 1978). The youngest Triassic zone here, the *marshi* zone, has three ammonoid species, two with single level records low in the zone, and only *Choristoceras marshi* is found throughout the zone. This, too, does not indicate a sudden TJB mass extinction of ammonoids. Thus, the change in ammonoids across the TJB is profound, but both global data and actual stratigraphic ranges indicate it took place as a series of extinction events spread across Norian and Rhaetian time, not as a single mass extinction at the TJB.

The evolutionary turnover of ammonoids across the TJB is an important change from diverse and morphologically complex forms (including various heteromorphs) to less diverse and morphologically simple forms (the psiloceratids). Guex (2001, 2006) argued that this kind of morphological change occurred in response to environmental stress, as had occurred at several other crisis points in the history of the Ammonoidea. The TJB was such a crisis in ammonoid history, but not a single mass extinction.

8.4.4 Reef builders

The scleractinian corals, important reef builders during the Triassic, suffered a marked decline at the end of the Triassic that was followed by a "reef gap" during part of the Early Jurassic (Hettangian-early Sinemurian), after which corals re-diversified to become the dominant reef builders (Stanley 1988). Stanley (2001, p. 26) viewed this as a "rapid collapse" of reefs at the TJB, concluded it was "the result of a first-order mass extinction" and noted that "Jurassic recovery was slow."

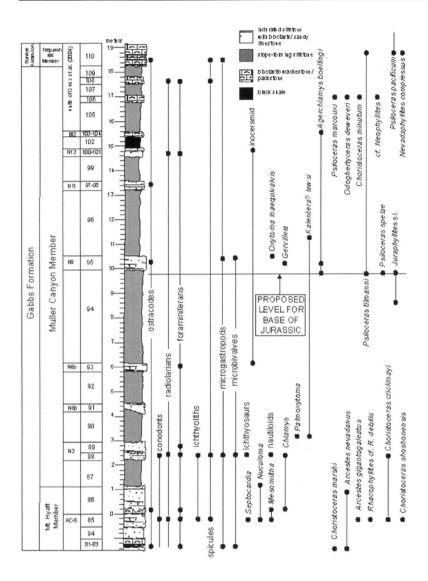

Fig. 5. The actual stratigraphic ranges of all known taxa (including ammonoids and bivalves) across the TJB in the Ferguson Hill section near New York Canyon, Nevada (modified from Lucas et al., 2007a). The TJB is placed here at the lowest occurrence of *Psiloceras tilmanni*

The extinctions in the reef community at the end of the Triassic are best documented in Tethys, where the reef ecosystem collapsed at the end of the Triassic, carbonate sedimentation nearly ceased, and earliest Jurassic reefal facies are rare. Earliest Jurassic reefs that are known (particularly in Morocco) are carbonate mounds produced by spongiomorphs and algae (e.g. Flügel 1975). However, coral Lazarus taxa have been discovered in Early Jurassic suspect terranes of western North America, indicating the persistence of at least some corals in Panthalassan refugia (on oceanic islands) during the earliest Jurassic reef gap (Stanley and Beauvais 1994).

Hallam and Goodfellow (1990) argued that sea level change caused the collapse of the reef system, with significant extinctions of calcisponges and scleractinian corals at the TJB. They discounted the possibility of a major drop in productivity as an explanation for the facies change from platform carbonates to siliciclastics. Indeed, a distinct lithofacies change occurs at or near the TJB in many sections, particularly in the Tethyan realm, where facies changes suggest an interval of regression followed by rapid transgression (Hallam and Wignall 1999). At the TJB section in western Austria, for example, a shallowing-upward trend from subtidal carbonates to red mudstones, interpreted as mudflat deposits, is succeeded by deeper water thin-bedded marl and dark limestone (McRoberts et al. 1997). The boundary in parts of the Austrian Alps displays karstification, suggesting a brief interval of emergence. In the Lombardian Alps the TJB is placed (palynologically) in the uppermost Zu Limestone at a flooding surface that marks the transition from mixed siliclastic-carbonate sedimentation to subtidal micrite deposition (Cirilli et al. 2003). Thus, a change in bathymetry resulted in the extirpation of reefs in Tethys, which in large part caused the cessation of carbonate sedimentation. However, the evidence that this was a global event is lacking, and it can be viewed as a regional (circum-Tethyan), not global, extinction driven by sea level changes (Tanner et al. 2004).

Kiessling et al.'s (1999) and Kiessling's (2001) global compilation indicates that the decline of reefs began during the Carnian and that the TJB corresponds to the loss of reefs concentrated around 30°N latitude. Nevertheless, this article has been cited as documenting a TJB mass extinction of reef organisms (e.g. Pálfy 2003). However, the timescale used in Kiessling's compilation is very coarse (it is only divided into Ladinian-Norian-Pliensbachian) and shows a steady decline in diversity throughout this time interval to reach a diversity low in the Middle Jurassic (Bajocian/Bathonian). This did not deter Kiessling et al. (1999), however, from identifying a major extinction of reefs at the TJB.

Flügel and Senowbari-Daryan (2001) drew a broader picture of Triassic reef evolution (also see Flügel, 2002). Thus, after the end-Permian mass extinction of the Paleozoic corals, there was a "reef gap" during the Early Triassic (Fig. 6). In the Middle Triassic (Anisian), reef-building resumed with the first appearance of scleractinian corals. The primary Middle Triassic reef builders, however, were stromatoporoids, calcisponges and encrusting algae. This continued into the early Carnian when there was a major evolutionary turnover in reefs that led to scleractinian dominance of reefs by Norian time.

Indeed, Flügel and Senowbari-Daryan (2001) referred to Norian-Rhaetian reefs as the "dawn of modern reefs" because they were characterized by abundant, highly calcified, sessile, gregarious and high-growing corals, as are modern reefs. However, Flügel and Senowbari-Daryan (2001) drew attention to a dramatic extinction of coral species at the TJB, with their analysis indicating that only 1% of Triassic coral species (14 of 321) and 8.6% of sphinctozoid coralline sponge genera (5 of 58) surviving the end of the Triassic. They also noted that the most successful of the Late Triassic corals, the distichophyllids, became totally extinct at the end of the Triassic (also see Roniewicz and Morycova 1989).

Coral reefs are extremely rare in the Hettangian-early Sinemurian, but Lazarus taxa from oceanic island refugia in Panthalassa (Stanley and Beauvais 1994) undermine the case for a simple global mass extinction of corals at the TJB. Furthermore, it is important to stress that by Pliensbachian time the reef ecosystem was well on its way to recovery. The Jurassic reefs continued to be dominated by scleractinian corals, so the disruption of the reef ecosystem was not permanent.

Beauvais (1984) stressed the endemism of scleractinian species during the Liassic, raising the possibility that the apparent TJB extinction of these organisms may be heavily influenced by (Tethyan?) sampling biases. Thus, a sudden extinction of reef organisms at the TJB is well documented in Tethys and reflects a regional change in bathymetry, but not necessarily a global mass extinction of reef organisms.

8.4.5 Conodonts

The Conodonta (a phylum or subphylum) is usually identified as one of the most significant groups to have suffered complete extinction at the end of the Triassic. This is misleading. Detailed reviews of conodont extinctions emphasize that conodonts suffered high rates of extinction and low rates of origination throughout the Triassic (e.g. Clark 1983, 1986, 1991; Sweet 1988; Kozur and Mock 1991; Aldridge and Smith 1993; De Renzi et al. 1996). During the Triassic, conodont diversity was highest during the Ladinian, and the ensuing Late Triassic saw a stepwise decline in this diversity as extinction rates were relatively high and origination rates were low. The single largest Late Triassic extinction took place during the Carnian (at the Julian-Tuvalian boundary), when nearly all platform conodonts disappeared (Kozur and Mock 1991). Diversity recovered through the Norian to decline again into the Rhaetian. Within the Rhaetian, nearly all conodont taxa disappeared before the TJB, with only one or two taxa found in youngest Rhaetian conodont assemblages (Mostler et al. 1978; Kozur and Mock 1991; Orchard 1991, 2003; Orchard et al. 2007).

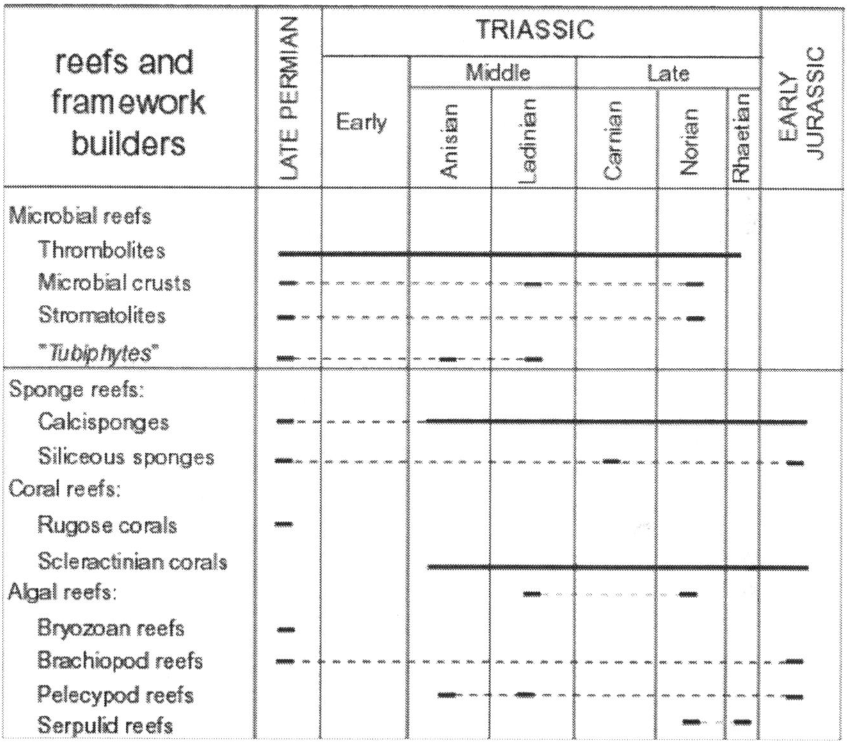

reefs and framework builders	LATE PERMIAN	TRIASSIC						EARLY JURASSIC
		Early	Middle		Late			
			Anisian	Ladinian	Carnian	Norian	Rhaetian	
Microbial reefs								
Thrombolites		────────────────────────						
Microbial crusts		─ ─ ─ ─ ─ ─ ─ ─ ─ ─ ─ ─ ─ ─						
Stromatolites		─ ─ ─ ─ ─ ─ ─ ─ ─ ─ ─ ─ ─ ─						
"*Tubiphytes*"		─ ─ ─ ─ ─ ─ ─ ─ ─ ─						
Sponge reefs:								
Calcisponges		─ ─ ────────────────						
Siliceous sponges		─ ─ ─ ─ ─ ─ ─ ─ ─ ─ ─ ─ ─ ─						
Coral reefs:								
Rugose corals		─						
Scleractinian corals				────────────────				
Algal reefs:					─── ──			
Bryozoan reefs		─						
Brachiopod reefs		─ ─ ─ ─ ─ ─ ─ ─ ─ ─ ─ ─ ─ ─						
Pelecypod reefs		─ ─ ─ ─ ─ ─ ─ ─ ─ ─ ─ ─ ─ ─						
Serpulid reefs							── ──	

Fig. 6. Temporal distribution of major Triassic reef types as characterized by the principal reef-building groups (after Flügel and Senowbari-Daryan 2001)

Furthermore, Pálfy et al. (2007) recently presented an extension of the age of the youngest Conodonta (also see Kozur 1993). From Csövár in Hungary they reported the conodont "*Neohindeodella*" *detrei* from a horizon stratigraphically above an indeterminate psiloceratid ammonoid that they took to indicate an Early Jurassic age. This appears to be the first documentation of a Jurassic conodont, and thus eliminates the final extinction of conodonts as an end-Triassic event (Fig. 2).

8.4.6 Land plants

An extensive literature documents the lack of a major extinction/turnover of the terrestrial macroflora and microflora at the TJB (e.g. Orbell 1973; Schuurman 1979; Pedersen and Lund 1980; Fisher and Dunay 1981; Brugman 1983; Niklas et al. 1983; Knoll 1984; Ash 1986; Traverse 1988; Edwards 1993; Cleal 1993a, b; Kelber 1998; Hallam 2002; Tanner et al. 2004; Lucas and Tanner 2004, 2007b; Galli et al. 2005; Ruckwied et al. 2006; Kuerschner et al. 2007). Thus, for example, Ash (1986) reviewed the global record of megafossil plants and concluded that changes across the TJB boundary primarily involved seed ferns, in particular, the loss of the families Glossopteridaceae, Peltaspermaceae, and Corystospermaceae (also see Traverse 1988). This accords well with the global compilations at the species and family levels that show no substantial extinction at the TJB (Niklas et al. 1983; Knoll 1984; Edwards 1993; Cleal 1993a, b).

An exception to prevailing thought was McElwain et al. (1999), who claimed a significant macrofloral extinction at the TJB in East Greenland, although previously, Harris (1937) and Pedersen and Lund (1980) interpreted the same data to indicate species range truncations at a depositional hiatus. The TJB in East Greenland is marked by the transition from the *Lepidopteris* floral zone to the *Thaumatopteris* floral zone, with few species shared by both zones. The former is characterized by the presence of palynomorphs including *Rhaetipollis*, while the latter contains *Heliosporites* (Pedersen and Lund 1980), and although extinction of some species across the transition between the two zones is evident, many species occur in both zones. Thus, no catastrophic extinction is documented and, at most, the floral turnover in East Greenland is nothing more than a local event, as no similar event is documented elsewhere (Hallam and Wignall 1997; Tanner et al. 2004).

The palynological record provides no evidence for mass extinction at the TJB. Thus, Fisher and Dunay (1981) demonstrated that a significant proportion of the *Rhaetipollis germanicus* assemblage that defines the Rhaetian in Europe (Orbell 1973; Schuurman 1979) persists in lowermost Jurassic strata. Indeed, a study of the British Rhaeto-Liassic by Orbell (1973) found that of 22 palynomorphs identified in the *Rhaetipollis* Zone, only 8 disappeared completely in the overlying *Heliosporites* Zone. These authors, as well as Brugman (1983) and Traverse (1988), have concluded that floral turnover across the TJB was gradual, not abrupt. Kelber (1998) also described the megaflora and palynoflora for Central Europe in a single unit he termed "Rhaeto-Liassic," and concluded there was no serious disruption or decline in plant diversity across the TJB.

More recently, Kuerschner et al. (2007) documented in detail the transitional nature of the change in palynomorphs in the Kössen and Kendelbach formations in the Tiefengraben section (Northern Calcareous Alps). They describe a *Rhaetipollis-Limbosporites* zone, correlative with the *C. marshii* ammonoid zone, in which *Corollina* (both *torosa* and *meyeriana*) is abundant. The overlying *Rhaetipollis-Porcellispora* zone, in which *R. germanicus* disappears near the top, contains a *Corollina* peak, but also contains Triassic foraminifera. The succeeding *Trachysporites-Porcellispora* zone marks a decline in *Corollina* and the disappearance of *Ovalipollis pseudolatus*. The overlying *Trachysporites-Heliosporites* zone is characterized by the maximum abundance of *H. reissingeri*. These authors suggest that the Triassic-Jurassic boundary in this section can be placed within the *Trachysporites-Porcellispora* zone, which corresponds to Schuurman's (1977, 1979) Phase 4, or between it and the *Trachysporites-Heliosporites* zone, which corresponds to Phase 5 of Schuurman (1977, 1979).

Nevertheless, profound palynomorph extinction at the TJB has been argued from the Newark Supergroup record in eastern North America (Olsen and Sues 1986; Olsen et al. 1990; Fowell and Olsen 1993; Olsen et al. 2002a, b) (Fig. 7). Notably, the palynomorph taxa used to define the TJB in the European sections (*Rhaetipollis germanicus* and *Heliosporites reissingeri*: Orbell 1973) are not present in the Newark Supergroup basins, so placement of the palynological TJB in these basins was initially based on a graphic correlation of palynomorph records (Cornet 1977). More recent work identified the TJB in the Newark by a decrease in diversity of the pollen assemblage, defined by the loss of palynomorphs considered typical of the Late Triassic, and dominance by several species of the genus *Corollina*, especially *C. meyeriana* (Cornet and Olsen 1985; Olsen et al. 1990; Fowell and Olsen 1993; Fowell et al. 1994; Fowell and Traverse, 1995).

This change has either been equated to the TJB or, most recently, referred to as the "T-J palynofloral turnover" (Whiteside et al. 2007). But, as Kozur and Weems (2005: 33) well observed, "there are no age-diagnostic sporomorphs or other fossils to prove that this extinction event occurred at the Triassic-Jurassic boundary." Kuerschner et al. (2007) further concluded that the Newark palynological event most likely represents an older, potentially early Rhaetian event, a conclusion shared by Kozur and Weems (2005, 2007) and by Lucas and Tanner (2007b).Thus, the palynological turnover in the Newark preceded the TJB and was a regional event, not a global mass extinction. There is no evidence of a global mass extinction of land plants at the TJB.

8.4.7 Tetrapods

The idea of a substantial nonmarine tetrapod (amphibian and reptile) extinction at the TJB began with Colbert (1949, 1958), and has been more recently advocated by Olsen et al. (1987, 1990, 2002a, b), largely based on the tetrapod fossil record of the Newark Supergroup (eastern North America). Weems (1992), Benton (1994), Lucas (1994), Tanner et al. (2004) and Lucas and Tanner (2004, 2007b) rejected this conclusion, arguing against building a case for extinction on the very incomplete vertebrate fossil record of the Newark Supergroup.

Indeed, Huber et al. (1993) plotted the stratigraphic ranges of all Triassic tetrapod taxa known from the Newark Supergroup, and only three body-fossil taxa (indeterminate phytosaurs, the procolophonid *Hypsognathus* and the sphenodontid *Sigmala*) actually are found in the youngest Triassic strata immediately below the oldest basalt sheet of the Central Atlantic Magmatic Province (CAMP) (Fig. 7). The last decade and a half of collecting has not changed that, and only a few, fragmentary tetrapod fossils are known from the Newark extrusive zone (Fig. 7) and are not age diagnostic (Lucas and Huber 2003). The McCoy Brook Formation, which overlies the only CAMP basalt of the Fundy basin in Nova Scotia, yields a tetrapod assemblage generally considered Early Jurassic in age by vertebrate paleontologists (Olsen et al. 1987; Shubin et al. 1994; Lucas 1998; Lucas and Huber 2003; Lucas and Tanner 2007a, b), though it could straddle the marine-defined TJB.

The Newark body fossil record of tetrapods is thus sparse across the TJB and therefore inadequate to evaluate a possible TJB tetrapod extinction, and the direct correlation of such an extinction (if it exists) to the marine TJB has not been demonstrated (Lucas and Tanner 2007b). Indeed, the most substantial extinction of tetrapods across the TJB is the crurotarsan extinction, which occurs below the lowest CAMP basalt in the Newark Supergroup (Fig. 7). This is the extinction of phytosaurs, aetosaurs and rauisuchians ("thecodonts") that has long represented the bulk of the supposed terminal Triassic tetrapod extinction.

Because the Newark Supergroup body fossil record of tetrapods is inadequate to demonstrate a mass extinction of tetrapods at the TJB, the tetrapod footprint record in the Newark Supergroup has been used instead (e.g. Olsen and Sues 1986; Szajna and Silvestri 1996; Olsen et al. 2002a,b). However, detailed stratigraphic study of the Newark footprint record indicates the disappearance of only four ichnogenera and the appearance of only two ichnogenera below the lowest CAMP basalt sheet, with four ichnogenera continuing through this boundary; this does not qualify as a sudden mass extinction.

Part of the footprint change in the Newark Supergroup below the oldest CAMP basalt is the lowest occurrence of the theropod footprint ichnogenus *Eubrontes* (as defined by Olsen et al. 1998, i.e., tridactyl theropod pes tracks longer than 28 cm). The idea that the lowest occurrence of *Eubrontes* is the base of the Jurassic began with Olsen and Galton (1984), who advocated this datum based on palynostratigraphy, not on the stratigraphic distribution of the footprints, themselves. Olsen et al. (2002a, b) further argued that the sudden appearance of *Eubrontes*, made by a *Dilophosaurus*-like theropod, in the "earliest Jurassic" strata of the Newark Supergroup, indicates a dramatic size increase in theropod dinosaurs at the TJB. They interpreted this as the result of a rapid (thousands of years) evolutionary response by the theropod survivors of a mass extinction and referred to it as "ecological release" (Olsen et al. 2002a, p. 1307). They admitted, however, that this hypothesis can be invalidated by the description of *Dilophosaurus*-sized theropods or diagnostic *Eubrontes* tracks in verifiably Triassic-age strata.

Indeed, tracks of large theropod dinosaurs assigned to *Eubrontes* (or its synonym *Kayentapus*) are known from the Triassic of Australia, Africa (Lesotho), Europe (Great Britain, France, Germany, Poland-Slovakia, Scania) and eastern Greenland, invalidating the "ecological release" hypothesis (Lucas et al. 2006). A detailed review of these records indicates Carnian, Norian and Rhaetian occurrences of tracks that meet the definition of *Eubrontes* established by Olsen et al. (1998). Also, theropods large enough to have made at least some *Eubrontes*-size tracks are known from the Late Triassic body-fossil record. The sudden abundance of these tracks in the Newark Supergroup cannot be explained simply by rapid evolution of small theropods to large size following a mass extinction. The concept of a sudden appearance of *Eubrontes* tracks due to "ecological release" at the TJB thus proposed by Olsen et al. (2002a, b) can be abandoned. Furthermore, tetrapod footprints do not provide a basis for precise placement of the TJB in the Newark Supergroup. Most *Eubrontes* tracks are Jurassic, but many are clearly Triassic (Lucas et al. 2006).

A few body fossil taxa of tetrapods and a few ichnogenera do seem to be restricted to either Triassic or Jurassic strata (Lucas and Tanner 2007b). Thus, no crurotarsan (aetosaur, phytosaur and rauisuchian) body fossil is demonstrably Jurassic, so the presence of crurotarsan fossils still can be relied on to indicate a Triassic age. Therefore, the stratigraphically highest crurotarsan tracks (usually referred to *Brachychirotherium*, and largely thought to be the tracks of rauisuchians) are also apparently of Triassic age.

Colbert (1958) believed that the temnospondyl amphibians, a significant component of late Paleozoic and Early-Middle Triassic tetrapod assemblages, underwent complete extinction at the TJB. However, more recent discoveries have invalidated that conclusion.

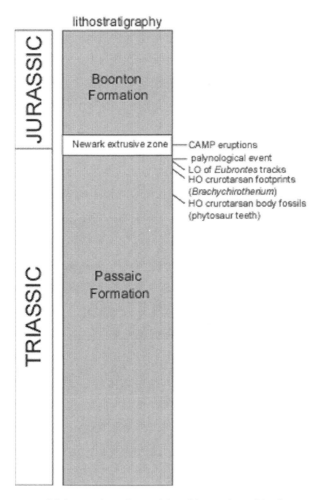

Fig. 7. Summary of lithostratigraphy and key biostratigraphic datum points of the Triassic-Jurassic boundary interval of the Newark Supergroup in the Newark basin of New Jersey-Pennsylvania, USA

Thus, Milner (1993, 1994) demonstrated a less pronounced extinction of temnospondyl amphibians, with only one family extinct at the end of the Triassic (plagiosaurids); he showed the disappearance of the capitosaurids, metoposaurids and latiscopids at the Norian-Rhaetian boundary. Moreover, these temnospondyls are only a minor component of Late Triassic tetrapod assemblages, being of low diversity and relatively small numbers in many samples (e.g. Hunt 1993). Temnospondyl extinction thus largely preceded the Rhaetian.

The global compilation of reptile families by Benton (1993) lists the extinction of 11 terrestrial reptile families at the TJB: Proganochelyidae, Kuehneosauridae, Pachystropheidae, Trilophosauridae, Phytosauridae, Stagonolepididae, Rauisuchidae, Ornithosuchidae, Saltoposuchidae, Thecodontosauridae and Traversodontidae. However, only two of these families, Phytosauridae and Procolophonidae, have well established Rhaetian records (Lucas 1994), especially given that new data indicate that the uppermost Chinle Group in the western United States (which has the most extensive Late Triassic vertebrate fossil record known: Lucas 1997) is pre-Rhaetian (Lucas and Tanner 2007a; Lucas et al. 2007). There is thus no evidence that most of the tetrapod families that disappeared during the Late Triassic were present during the Rhaetian; they apparently became extinct sometime earlier, during the Norian.

Tetrapod extinctions close to the TJB thus are the extinctions of some crurotarsans, particularly the phyosaurs and (based on tracks) possibly the rauisuchians. The complex and prolonged turnover from crurotarsan- to dinosaur-dominated tetrapod assemblages began in the Late Triassic.

8.5 Ecological severity

McGhee et al. (2004) made the valuable point that not only should mass extinctions be evaluated in terms of biodiversity crises, but also in terms of their ecological severity. In their scheme of ecological severity, they evaluated the marine TJB extinction as category IIa and the continental TJB extinction as category I or IIa. Category I means that ecosystems before the extinction were replaced by new ecosystems post-extinction, whereas category IIa means that the extinctions caused permanent loss of major ecosystem components.

McGhee et al. (2004: 291) rated the TJB marine extinction as category IIa because of the "virtual elimination of the global reef component of marine ecosystems." However, as discussed above, this disruption was not demonstrably global and it was demonstrably temporary (Fig. 6). Therefore, we downgrade the TJB marine extinction to category IIb in their classification, which means that the disruption was temporary; i.e., the reef ecosystem re-established itself after a hiatus.

McGhee et al. (2004: 293) considered characterizing the ecological severity of the nonmarine TJB extinction as "more problematic" than their characterization of the marine TJB extinction. Despite this, they concluded that the TJB transition involved a rapid ecological replacement of Triassic mammal-like reptiles and rhynchosaurs by dinosaurs. However, rhynchosaurs became extinct during the late Carnian (Hunt and Lucas 1991; Lucas et al. 2002) and dicynodonts were also extinct well before the end of the Triassic (Lucas and Wild 1995), unless a putative Cretaceous record (with problematic provenance) from Australia is verified (Thulborn and Turner 2003). The other principal group of Late Triassic mammal-like reptiles, the cynodonts, were of low diversity after the Carnian (Lucas and Hunt 1994). Dinosaurs appeared as body fossils in the Carnian and had begun to diversify substantially in some parts of Pangea by the late Norian (e.g. Hunt 1991). Thus, the ecological severity of the end-Triassic tetrapod extinction is relatively low (Category IIb on the McGhee et al. 2004 classification), and the plant extinctions don't look like they were ecologically severe either (see above). Clearly, there was some disruption of the terrestrial ecosystem across the TJB, but it was not severe.

One of the most significant paleoecological events of the Mesozoic was the Mesozoic marine revolution. This involved an increase in predation pressure (particularly in durivorous predators) and the turnover from marine benthic communities dominated by epifaunal (surface-dwelling) or semi-infaunal animals to a more infaunal benthos (e.g. Vermeij 1977, 1983; Harper and Skelton 1993). The Mesozoic marine revolution began in the Triassic but did not really accelerate until well into the Jurassic, during the Pliensbachian-Toarcian. Thus, the TJB has no clear relationship to the Mesozoic marine revolution.

8.6 Causes of the TJB extinctions

The end-Triassic drop in diversity was selective but notable for the rapid loss of specific marine taxa, such as ammonoids, conodonts, radiolarians and infaunal bivalves, suggesting physical processes that strongly affected ocean bioproductivity (Tanner et al. 2004). Moreover, the temporary loss of scleractinian corals and almost all calcareous nannoplankton has suggested to some a "calcification crisis" coincident with the above losses (Hautmann 2004; Schootbrugge et al. 2007). These biotic events coincide with a significant excursion in the carbon isotope composition of organic matter that has been identified in close proximity to the system boundary in numerous marine sections.

For example, the sections at St. Audrie's Bay, southwest England (Hesselbo et al. 2002, 2004), Csövár, Hungary (Pálfy et al. 2001), and Tiefengraben, Austria (Kuerschner et al. 2007) display negative $\delta^{13}C$

excursions of approximately 2.0 to 3.0 ‰. Consistently, these excursions begin below the highest occurrence (HO) of conodonts, supporting their correlatability. At the Kennecott Point section in the Queen Charlotte Islands, Canada, a negative $\delta^{13}C$ excursion (of approximately 1.5-2.0 ‰) spans the TJB, beginning immediately below the HO of Triassic ammonites and radiolarians, and continuing above the lowest occurrence (LO) of Jurassic radiolarians (Ward et al. 2001, 2004; Williford et al. 2007). In the New York Canyon section of Nevada, USA, a negative $\delta^{13}C$ excursion of similar magnitude (about 2.0 ‰) also begins just below the HO of conodonts, Triassic ammonoids (*Choristoceras crickmayi* and *Arcestes* spp.) and characteristic Triassic bivalves (Guex et al. 2004; Lucas et al. 2007a).

Most significantly, global end-Triassic events included the initiation of the CAMP eruptions, with a volume of >2 x 10^6 km^3. The CAMP eruptions likely emitted a total of 2.3 x 10^{18} g of sulfur as SO_2 (McHone 2003). The environmental effects of large sulfur emissions during such prolonged flood basalt eruptions are not clear, but the formation of H_2SO_4 aerosols is known to increase atmospheric opacity and result in reduced short-wave radiant heating, causing global cooling (Sigurdsson 1990). Individual CAMP eruptive pulses were likely of sufficient size to inject 10^{16} g or more of sulfur into the atmosphere and cause immediate cooling episodes of as much as 6°C (Sigurdsson 1990). Cooling of this magnitude is consistent with some interpretations of the palynological record (Hubbard and Boulter 2000).

Outgassing during individual CAMP eruptions also would have added large volumes of CO_2 to the atmosphere. Calculations based on CAMP compositional data suggest a total release of 5.19 x 10^{18} g of CO_2 (McHone 2003). This addition to the CO_2 rich atmosphere of the early Mesozoic would have had little immediate effect. However, the longer residence time of CO_2 in the atmosphere allowed pCO_2 to increase continually through the duration of the eruptions by several hundred ppm, causing greenhouse warming of ~2°C or more (Beerling and Berner 2002) that would gradually offset and finally replace radiative cooling as acid aerosols were removed from the atmosphere. This greenhouse state would have persisted for an extended interval (i.e., several hundred thousand years or more) after the cessation of the eruptions.

The environmental disruption at the TJB likely included the effects of dramatic temperature fluctuations resulting from intense radiative cooling, potentially exceeding 6°C, followed subsequently by greenhouse warming of 2°C or more. Fluorine and chlorine volatile emissions during the CAMP eruptions also have been suggested as an environmental consequence of the eruptions that contributed to Late Triassic extinctions (Guex et al.

2004). Under modern conditions, the ocean's large buffering capacity would prevent a dramatic change in alkalinity except under the most extreme conditions (Berner and Beerling 2007). Hence, a "calcification crisis" is not likely. Rather, we envision a temporary loss of phytoplankton productivity in surface waters that had "trickle-down" effects through the trophic system. Most radiolarians, for example, live in surface waters (<100 m water depth), and thus would have been profoundly affected by the loss of primary productivity. Higher level consumers felt the ecological pressure of this loss less strongly.

A negative $\delta^{13}C$ excursion for organic carbon in terrestrial environments also has been claimed for nonmarine strata encompassing the TJB, but these data are problematic (McElwain et al. 1999; Hesselbo et al. 2002). The presumption is that the marine excursion resulted from a drastic alteration of the $\delta^{13}C$ of the global CO_2 reservoir that similarly was recorded by vascular plants. However, published isotope analyses of plant macrofossils for TJB sections in Scania fail to exhibit this excursion in any fashion (McElwain et al. 1999). Data from Greenland display an apparent trend that appears to correlate with the marine data, although there is significant intersample $\delta^{13}C$ variability and therefore a lack of the consistency that is displayed in the marine record (McElwain et al. 1999; Hesselbo et al. 2002). Notably, terrestrial organic matter displays significant interspecific variations in isotopic composition due to variations in the organic composition. Furthermore, variations in the isotopic composition of plants may result from environmental factors other than the $\delta^{13}C$ of the atmosphere. Combined with a lack of data for significant extinction in the terrestrial realm, we thus find little evidence that events at the TJB had a great impact on the land-based biota.

8.7 Late Triassic extinction events

The Late Triassic was a time of elevated extinction rates and low origination rates in many biotic groups (e.g. Banmbach et al. 2004; Kiessling et al. 2007). Thus, as noted by many workers, the Late Triassic was a time interval marked by a series of discrete extinction events (Fig. 8). One of the most dramatic was the "Carnian crisis" at about the early-middle Carnian boundary, which included major extinctions of crinoids (especially the Encrinidae), echinoids, some bivalves (scallops), bryozoans, ammonoids, conodonts and a major change in the reef ecosystem (see above) in the seas (e.g. Schäfer and Fois 1987; Johnson and Simms 1989; Hallam 1995; Flügel 2002; Hornung et al. 2007). On land, plant and vertebrate extinctions seem less dramatic within the Carnian. Most workers envision a Carnian humid phase (or "pluvial") as a possible cause of this, at least in Europe (e.g. Simms and Ruffell 1989, 1990; Rigo et al. 2007; Hornung et al. 2007), though this has been disputed by some (e.g. Visscher et al. 1994).

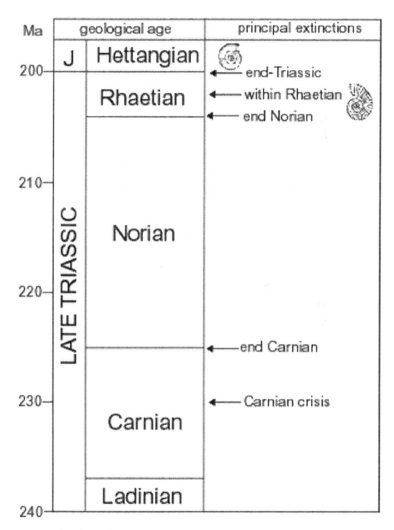

Fig. 8. Late Triassic extinction events

There is also a Carnian-Norian boundary extinction event in the terrestrial tetrapod record, with some evolutionary turnover across the Carnian-Norian boundary (Benton 1986, 1991; Lucas 1994). However, the case for an extinction of marine reptiles at this boundary (Benton 1986) is not confirmed by a more detailed analysis of Triassic marine reptile diversity (Bardet, 1995). Nevertheless, in the sea, there is extinction of conodonts, ammonoids and some bivalves (especially pectinids) at the Carnian-Norian boundary (Johnson and Simms 1989).

Within the Norian, there were several extinction events culminated by an extinction at the end of the Norian that had a particularly profound affect on marine bivalves and ammonoids (see above). Further significant extinctions in these groups (and of conodonts) took place within the Rhaetian.

Two hundred years of fossil collecting failed to document a global mass extinction at the TJB, yet 25 years of literature compilation and the compiled correlation effect did. The idea of a single mass extinction at the TJB has led to a search for the cause of the "mass extinction" and drawn attention away from what were actually a series of extinctions that took place throughout the Late Triassic (Fig. 8). Research should now focus on these multiple extinctions and their causes, not on a single extinction event. Perhaps the most interesting question not yet addressed by most researchers is: why this prolonged (at least 20 million years) interval of elevated extinction rates occurred during the Late Triassic?

8.8 Acknowledgments

We are grateful to numerous colleagues whose ideas and work have influenced this article. In particular, we thank Jean Guex, Tony Hallam, Steve Hesselbo, Heinz Kozur, Wolfram Kuerschner, Leo Krystyn, Chris McRoberts, Paul Olsen, Josef Pálfy and Geoff Warrington, not all of whom agree with our conclusions, but all of whom have contributed substantially to our understanding of events across the TJB.

References

Aldridge RJ, Smith MP (1993) Conodonta. In Benton MJ (ed) The fossil record 2, Chapman & Hall, London pp 563-572
Allasinaz A (1992) The Late Triassic-Hettangian bivalve turnover in Lombardy (Southern Alps). Rivista Italiana Paleontogia Stratigrafia 97: 431-454.
Ash S (1986) Fossil plants and the Triassic-Jurassic boundary. In Padian K (ed) The beginning of the age of dinosaurs. Cambridge University Press, Cambridge pp 21-30

Bambach RK, Knoll AH Wang SC (2004) Origination, extinction, and mass depletion of marine diversity. Paleobiology 20: 522-542

Bardet N (1995) Evolution et extinction des reptiles marins au cours du Mesozoique. Palaeovertebrata 24: 177-283

Barras CG, Twitchett RJ (2007) Response of the marine infauna to Triassic-Jurassic environmental change: Ichnological data from southern England. Palaeogeography, Palaeoclimatology, Palaeoecology 244: 223-241

Beauvais L (1984) Evolution and diversification of Jurassic Scleractinia. Palaeontographia Americana 54: 219-224

Beerling DJ, Berner RA (2002) Biogeochemical constraints on the Triassic–Jurassic boundary carbon cycle event. Global Biogeochemical Cycles 16: 101-113

Benton MJ (1986) More than one event in the Late Triassic mass extinction. Nature 321: 857-861

Benton MJ (1991) What really happened in the Late Triassic? Historical Biology 5: 263-278

Benton MJ (1993) Reptilia. In Benton MJ (ed) The fossil record 2. Chapman & Hall, London, 681-715

Benton MJ (1994) Late Triassic to Middle Jurassic extinctions among continental tetrapods: testing the pattern. In Fraser NC, Sues H-D (eds) In the shadow of the sinosaurs. Cambridge University Press, Cambridge, 366-397

Berner RA, Beerling DJ (2007) Volcanic degassing necessary to produce a $CaCO_3$ undersaturated ocean at the Triassic–Jurassic boundary. Palaeogeography, Palaeoclimatology, Palaeoecology 244: 368-373

Blome CD (1986) Paleogeographic significance of Upper Triassic and Lower Jurassic Radiolaria from Cordilleran terranes. Proceedings North American Paleontological Convention 4: A5

Blome CD, Hull DM, Pessagno Jr EA, Reed KM (1995) Meozoic Radiolaria. In Blome CD Whalen PM Reed KM (eds) Siliceous microfossils, short courses in paleontology 8. The Paleontological Society 31-60

Brugman WA (1983) Permian-Triassic palynology. State University Utrecht, Utrecht 121 pp

Carpenter K (1997) A giant coelophysoid (Ceratosauria) theropod from the Upper Triassic of New Mexico, USA. Neues Jahrbuch für Geologie und Paläontologie Abhandlungen 205: 189-208

Carter ES (1993) Biochronology and paleontology of uppermost Triassic(Rhaetian) radiolarians, Queen Charlotte Islands, British Columbia,Canada. Mémoires de Géologie, Lausanne 11 (175 pp, 21 pls)

Carter ES (1994) Evolutionary trends in latest Norian through Hettangian radiolarians from the Queen Charlotte Islands, British Columbia. Geobios Mémoire Spécial 17: 111-119

Carter ES (2007) Global distribution of Rhaetian radiolarian faunas and their contribution to the definition of the Triassic-Jurassic boundary. New Mexico Museum of Natural History and Science Bulletin 41: 27-31

Carter E, Hori R (2005) Global correlation of the radiolarian faunal change across the Triassic-Jurassic boundary. Canadian Journal of Earth Sciences 42: 777-790

Cirilli S, Galli MT, Jadoul F (2003) Carbonate platform evolution and sequence stratigraphy at Triassic/Jurassic boundary in the Western Southern Alps of Lombardy (Italy): an integrated approach of litho-palynofacies analysis. Geological Association of Canada, Vancouver 2003 Meeting, Abstracts Volume: 28

Clark DL (1980) Rise and fall of Triassic conodonts. American Association of Petroleum Geologists Bulletin 64: 691

Clark DL (1981) Extinction of Triassic conodonts. Geologische Bundesanstalt Abhandlungen 35: 193-195

Clark DL (1983) Extinction of conodonts. Journal of Paleontology 57: 652-661.

Cleal CJ (1993a) Pteridophyta. In Benton MJ (ed) The fossil record 2. Chapman & Hall, London, 779-794

Cleal CJ (1993) Gymnospermophyta. In Benton M J (ed) The fossil record 2. Chapman & Hall, London pp 795-808

Colbert EH (1949) Progressive adaptations as seen in the fossil record. In Jepsen GL, Mayr E, Simpson GG (eds) Genetics, paleontology and evolution. Princeton University Press, Princeton, 390-402

Colbert EH (1958) Triassic tetrapod extinction at the end of the Triassic Period. Proceedings National Academy Science USA 44: 973-977

Cope JCW, Getty TA, Howarth MK, Morton N Torrens HS (1980) A correlation of Jurassic rocks in the British Isles; Part one, Introduction and Lower Jurassic: Geological Society (London), Special Report 14

Cornet B (1977) The palynostratigraphy and age of the Newark Supergroup. Ph.D. Thesis. Pennsylvania State University, University Park, PA, 505 pp

Cornet B, Olsen PE (1985) A summary of the biostratigraphy of the Newark Supergroup of eastern North America with comments on provinciality. In Weber R (ed) III Congreso Latinoamericano de Paleontolia Mexico, Simposio Sobre Floras del Triasico Tardio, su Fitogeografia y Palecologia, Memoria. UNAM Instituto de Geologia, Mexico City pp. 67-81

Dagys AS, Dagys AA (1994) Global correlation of the terminal Triassic. Mémoire de Géologie (Lausanne) 22: 25-34

De Renzi, M Budurov K, Sudar M (1996) The extinction of conodonts—in terms of discrete elements—at the Triassic-Jurassic boundary. Cuadernos de Geología Ibérica 20: 347-364

Dommergues J-L, Laurin B, Meister C (2001) The recovery and radiation of Early Jurassic ammonoids: Morphologic versus palaeobiogeographical patterns. Palaeogeography, Palaeoclimatology, Palaeoecology 165: 195-213

Dommergues J-L, Montuire S, Neige P (2002) Size patterns through time: The case of the early Jurassuc ammonite radiation. Paleobiology 28: 423-434

Donovan DT, Callomon JH, Howarth MK (1980) Classification of the Jurassic Ammonitina In House MJ, Senior JR (eds) The Ammonoidea: Systematics Association Special Volume 18: 101-155

Edwards D (1993) Bryophyta. In Benton MJ (ed) The fossil record 2. Chapman & Hall, London pp 775-778
Fisher MJ, Dunay RE (1981) Palynology and the Triassic/Jurassic boundary. Review of Palaeobotany and Palynology 34: 129-135
Flügel E. (1975) Fossile Hydrozoan—Kenntnisse und Probleme. Paläontologische Zeitschrift 49: 369-406
Flugel E. (2002) Triassic reef patterns. In Kiessling W, Flügel E, Golonka J (eds) Phanerozoic reef patterns SEPM Special Publication 72: 391-463
Flugel E, Senowbari-Daryan B (2001) Triassic reefs of the Tethys. In Stanley GD Jr (ed) The history and sedimentology of ancient reef systems. Kluwer Academic/Plenum Publishers, New York, 217-249
Flügel E, Stanley GD Jr (1984) Reorganization, development and evolution of post-Permian reefs and reef organisms. Palaeontographica Americana 54: 177-186
Fowell SJ, Olsen PE (1993) Time calibration of Triassic-Jurassic microfloral turnover, eastern North America. Tectonophysics 222: 361-369
Fowell SJ, Traverse A (1995) Palynology and age of the upper Blomidon Formation, Fundy basin, Nova Scotia. Review of Palaeobotany and Palynology 86: 211-233
Fowell SJ, Cornet B, Olsen PE (1994) Geologically rapid Late Triassic extinctions: palynological evidence from the Newark Supergroup. Geological Society of America Special Paper 288: 197-206
Furin S, Preto N, Rigo M, Roghi G, Gianolla P, Crowley JL, Bowring SA (2006) High-precision U-Pb zircon age from the Triassic of Italy: Implications for the Triassic time scale and the Carnian origin of calcareous nannoplankton and dinosaurs. Geology 34: 1009-1012
Gallet Y, Krystyn L, Marcoux J, Besse J (2007) New constraints on the end-Triassic (Upper Norian–Rhaetian) magnetostratigraphy. Earth and Planetary Science Letters 255: 458-470
Galli, MT, Jadoul F, Bernasconi SM, Weissert H (2005) Anomalies in global carbon cycling and extinction at the Triassic/Jurassic boundary: Evidence from a marine C-isotope record. Palaeogeography, Palaeoclimatology, Palaeoecology 161: 203-214
Gómez JJ, Goy A, Barrón E (2007) Events around the Triassic-Jurassic boundary in northern and eastern Spain A review. Palaeogeography, Palaeoclimatology, Palaeoecology 244: 89-110
Gradstein FM, Ogg JG, Smith AG, Bleeker W, Lourens LJ (2004) A new geologic time scale with special reference to the Precambrian and Neogene. Episodes 27, 83-100
Guex J (1982) Relations entre le genre *Psiloceras* et les Phylloceratida au voisinage de la limite Trias-Jurassique. Bulletin Géologie Lausanne 260: 47-51
Guex J (1987) Sur la phylogenèse des ammonites du Lias inférieur. Bulletin Géologie Lausanne 305: 455-469
Guex J (2001) Environmental stress and atavism in ammonoid evolution. Eclogae Geolog Helvetica 94: 321-328

Guex J (2006) Reinitialization of evolutionary clocks during sublethal environmental stress in some invertebrates. Earth and Planetary Science Letters 242: 240-253

Guex J Bartolini A Atudorei V Taylor D (2003) Two negative $\delta^{13}C_{org}$ excursions near the Triassic-Jurassic boundary in the New York Canyon area (Gabbs Valley Range, Nevada). Bulletin de Géologie Lausanne 360: 1-4

Guex J, Bartolini A, Atudorei V, Taylor D (2004) High-resolution ammonite and carbon isotope stratigraphy across the Triassic-Jurassic boundary at New York Canyon (Nevada). Earth and Planetary Science Letters 225: 29-41

Guex J, Bartolini A, Taylor D (2002) Discovery of *Neophyllites* (Ammonita, Cephalopoda, early Hettangian) in the New York Canyon sections (Gabbs Valley Range, Nevada) and discussion of the $\delta^{13}C$ negative anomalies located around the Triassic-Jurassic boundary. Bulletin Société Vaudoise Sciences Naturelles 88.2: 247-255

Hallam A (1981) The end-Triassic bivalve extinction event. Palaeogeography, Palaeoclimatology, Palaeoecology 35: 1-44

Hallam A (1995) Major bio-events in the Triassic and Jurassic. In Walliser OH (ed) Global events and event stratigraphy. Springer, Berlin, 265-283

Hallam A (2002) How catastrophic was the end-Triassic mass extinction? Lethaia 35: 147-157

Hallam A, Goodfellow WD (1990) Facies and geochemical evidence bearing on the end-Triassic disappearance of the Alpine reef ecosystem. Historical Biology 4: 131-138

Hallam A, Wignall PB (1997) Mass Extinctions and their Aftermath. Oxford University Press, Oxford, 320 pp

Hallam A, Wignall PB (1999) Mass extinctions and sea-level changes. Earth Science Reviews 48: 217-258

Hallam A, Wignall PB (2000) Facies changes across the Triassic-Jurassic boundary in Nevada, USA. Journal Geological Society, London, 157: 49-54

Hallam A, Wignall PB, Yin J, Riding JB (2000) An investigation into possible facies changes across the Triassic-Jurassic boundary in southern Tibet. Sedimentary Geology 137: 101-106

Harper EM, Skelton PW (1993) The Mesozoic marine revolution and epifaunal bivalves. Scripta Geological Special Issue 2: 127-153

Harris TM (1937) The fossil flora of Scoresby Sound East Greenland, Part 5: Stratigraphic relations of the plant beds. Meddelelser om Grønland 112:1-112

Hart MB, Williams CL (1993) Protozoa. In Benton MJ (ed) The fossil record 2. Chapman & Hall, London, 43-70

Hautmann M (2004) Effect of end-Triassic CO_2 maximum on carbonate sedimentation and marine mass extinction. Facies 50: 257-261

Hesselbo SP, Robinson SA, Surlyk F, Piasecki S (2002) Terrestrial and marine extinction at the Triassic-Jurassic boundary synchronized with major carbon-cycle perturbation: a link to initiation of massive volcanism? Geology 30: 251-254

Hesselbo SP, Robinson SA, Surlyk F (2004) Sea-level change and facies development across potential Triassic-Jurassic boundary horizons, SW Britain. Journal Geological Society of London 161: 365-379

Hori R (1992) Radiolarian biostratigraphy at the Triassic/Jurassic period boundary in bedded cherts from the Inuyama area, central Japan. J. Geoscience Osaka City University 35: 53-65

Hornung T, Brandner R, Krystyn L, Jaochimski MM, Keim L (2007) Multistratigraphic constrainst on the NW Tethyan "Carnian crisis" New Mexico Museum of Natural History and Science Bulletin 41: 59-67

Hounslow MW, Posen PE, Warrington G (2004) Magnetostratigraphy and biostratigraphy of the Upper Triassic and lowermost Jurassic succession, St. Audrie's Bay, UK. Palaeogeography, Palaeoclimatology, Palaeoecology 213: 331-358

House MR (1989) Ammonoid extinction events. Philosophical Transactions Royal Society of London B 325: 307-326

Hubbard RN, Boulter MC (2000) Phytogeography and paleoecology in western Europe and eastern Greenland near the Triassic-Jurassic boundary. Palaios 15: 120-131

Huene F von (1934) Ein neuer Coelurosaurier in der thüringischen Trias. Paläontologische Zeitschrift 16: 145-170

Hunt AP (1993) A revision of the Metoposauridae (Amphibia: Temnospondyli) of the Late Triassic with description of a new genus from the western United States. Museum Northern Arizona Bulletin 59: 67-97

Hunt AP Lucas SG (1991) A new rhynchosaur from West Texas (USA) and the biochronology of Late Triassic rhynchosaurs. Palaeontology 34: 191-198

Johnson LA, Simms MJ (1989) The timing and cause of Late Triassic marine invertebrate extinctions: evidence from scallops and crinoids. In Donovan SK (ed) Mass extinctions: Processes and evidence. Columbia University Press, New York pp. 174-194

Kelber K-P (1998) Phytostratigraphische aspekte der makrofloren des süddeutschen Keupers. Documenta Naturae 117: 89-115

Kennedy WJ (1977) Ammonite evolution. In Hallam A (ed) Patterns of evolution as illustrated in the fossil record. Elsevier, Amsterdam pp 251-304

Kidder DL, Erwin DH (2001) Secular distribution of biogenic silica through the Phanerozoic: comparison of silica-replaced fossils and bedded cherts at the series level. Journal of Geology 109: 509-522

Kiessling W (2001) Paleoclimatic significance of Phanerozoic reefs. Geology 29: 751-754

Kiessling W, Flügel E, Golonka J (1999) Paleoreef maps: Evaluation of a comprehensive database on Phanerozoic reefs. American Association of Petroleum Geologists Bulletin 83: 1552-1587

Kiessling W, Aberhan M Brenneis, B Wagner PJ (2007) Extinction trajectories of benthic organisms across the Triassic-Jurassic boundary. Palaeogeography, Palaeoclimatology, Palaeoecology 244: 201-222

Knoll AH (1984) Patterns of extinction in the fossil record of vascular plants. In Nitecki MH (ed) Extinction. University of Chicago Press, Chicago pp 21-68

Kozur H (1993) First evidence of Liassic in the vicinity of Csovar (Hungary) and its paleogeographic and paleotectonic significance. Jahrbuch Geologische Bundes –Anstalt 136: 89-98

Kozur H, Mock R (1991) New middle Carnian and Rhaetian conodonts from Hungary and the Alps. Stratigraphic importance and tectonic implications for the Buda Mountains and adjacent areas. Jahrbuch Geologische Bundes–Anstalt 134: 271-297

Kozur HW, Weems RE (2005) Conchostracan evidence for a late Rhaetian to early Hettangian age for the CAMP volcanic event in the Newark Supergroup, and a Sevatian (late Norian) age for the immediately underlying beds. Hallesches Jahrbuch Geowissenschaft B27: 21-51

Kozur HW, Weems RE (2007) Upper Triassic conchostracan biostratigraphy of the continental rift basins of eastern North America: Its importance for correlating Newark Superegroup events with the Germanic Basin and the international geologic time scale. New Mexico Museum of Natural History and Science Bulletin 41: 137-188

Krystyn L, Bouquerel H, Kuerschner W, Richoz S, Gallet Y (2007) Proposal for a candidate GSSP for the base of the Rhaetian Stage. New Mexico Museum of Natural History and Science Bulletin 41: 189-199.

Kuerschner WM, Bonis NR, Krystyn L (2007) Carbon-isotope stratigraphy and palynostratigraphy of the Triassic-Jurassic transition in the Tiefengraben section—Northern Calcareous Alps (Austria). Palaeogeography, Palaeoclimatology, Palaeoecology 244: 257-280

Lloyd AJ (1964) The Luxembourg Colloquium and the revision of the stages of the Jurassic System. Geological Magazine 101: 249-259

Longridge LM, Carter ES, Smith PL, Tipper HW (2007) Early Hettangian ammonites and radiolarians from the Queen Charlotte Islands, British Columbia and their bearing on the definition of the Triassic–Jurassic boundary. Palaeogeography, Palaeoclimatology, Palaeoecology 244: 142-169

Lucas SG (1994) Triassic tetrapod extinctions and the compiled correlation effect. Canadian society Petroleum Geologists Memoir 17: 869-875

Lucas SG (1997) Upper Triassic Chinle Group, western United States: A nonmarine standard for Late Triassic time. In Dickins JM, Yang Z, Yin H, Lucas SG, Acharyya SK (eds) Late Palaeozoic and early Mesozoic circum-Pacific events and their global correlation. Cambridge Univerity Press, Cambridge pp 209-228

Lucas SG (1998) Global Triassic tetrapod biostratigraphy and biochronology. Palaeogeography, Palaeoclimatology, Palaeoecology 143: 347-384

Lucas SG, Huber P (2003) Vertebrate biostratigraphy and biochronology of the nonmarine Late Triassic In LeTourneau PM, Olsen PE (eds) The great rift Valleys of Pangea in eastern North America. Volume 2. Sedimentology, stratigraphy, and paleontology. Columbia University Press, New York pp 143-191

Lucas SG, Hunt AP (1994) The chronology and paleobiogeography of mammalian origins. In Fraser NC, Sues H-D (eds) In the shadow of dinosaurs: Early Mesozoic tetrapods. Cambridge University Press, New York pp 335-351

Lucas SG, Tanner LH (2004) Late Triassic extinction events. Albertiana 31: 31-40

Lucas SG, Tanner LH (2007a) Tetrapod biostratigraphy and biochronology of the Triassic-Jurassic transition on the southern Colorado Plateau, USA. Palaeogeography, Palaeoclimatology, Palaeoecology 244: 242-256.

Lucas SG, Tanner LH (2007b) The nonmarine Triassic-Jurassic boundary in the Newark Supergroup of eastern North America. Earth Science Reviews, in press.

Lucas SG, Wild R (1995) A Middle Triassic dicynodont from Germany and the biochronology of Triassic dicynodonts. Stuttgarter Beiträge zur Naturkunde 220: 1-16.

Lucas SG, Guex J, Tanner LH (2006) Criterion for definition of the Triassic/Jurassic boundary: Volumina Jurassica 4: 291.

Lucas SG, Heckert AB, Hotton N III (2002) The rhynchosaur *Hyperodapedon* from the Upper Triassic of Wyoming and its global biochronological significance. New Mexico Museum of Natural History and Science Bulletin 21: 149-156.

Lucas SG, Guex J, Tanner LH, Taylor D, Kuerschner WM, Atudorei V, Bartolini A (2005) Definition of the Triassic-Jurassic boundary. Albertiana 32: 12-16.

Lucas SG, Hunt AP, Heckert AB, Spielmann JA (2007b) Global Triassic tetrapod biostratigraphy and biochronology: 2007 status. New Mexico Museum of Natural History and Science Bulletin 41: 229-240.

Lucas SG, Klein H, Lockley MG, Spielmann JA, Gierlinski G, Hunt AP, Tanner LH (2006) Triassic-Jurassic stratigraphic distribution of the theropod footprint ichnogenus *Eubrontes*. New Mexico Museum of Natural History and Science Bulletin 37: 86-93.

Lucas SG, Taylor DG, Guex J, Tanner LH, Krainer K (2007a) The proposed global stratotype section and point for the base of the Jurassic System in the New York Canyon area, Nevada, USA.New Mexico Museum of Natural History and Science Bulletin 40: 139-168.

Marshall C (2005) Comment on "Abrupt and gradual extinction among Late Permian land vertebrates in the Karoo basin, South Africa." Science 308: 1413-1414.

Maubeuge P–L (1964) Résolutions du colloque. In Maubeuge, P–L (ed) Colloque du Jurassique à Luxembourg 1962. Ministère des Arts et des Sciences, Luxembourg, 77-80

McElwain JC, Beerling DJ, Woodward FI (1999) Fossil plants and global warming at the Triassic-Jurassic boundary. Science 285: 1386-1390

McGhee GR Jr, Sheehan PM, Bottjer DJ, Droser ML (2004) Ecological ranking of Phanerozoic biodiversity crises: Ecological and taxonomic severities are decoupled. Palaeogeography, Palaeoclimatology, Palaeoecology 211: 289-297

McHone JG (2003) Volatile emissions from central Atlantic magmatic province basalts: mass assumptions and environmental consequences. In Hames WE,

McHone JG, Renne PR, Ruppel C (eds) The Central Atlantic Magmatic Province: Perspectives from the rifted fragments of Pangea, AGU, Washington DC Monograph 136: 241-254

McRoberts CA (1994) The Triassic-Jurassic ecostratigraphic transition in the Lombardian Alps, Italy. Palaeogeography, Palaeoclimatology, Palaeoecology 110: 145-166

McRoberts CA (2007) Diversity dynamics and evolutionary ecology of Middle and Late Triassic halobiid and monotid bivalves. New Mexico Museum of Natural History and Science Bulletin 41: 272

McRoberts CA, Newton CR (1995) Selective extinction among end-Triassic European bivalves. Geology 23: 102-104

McRoberts CA, Furrer H, Jones DS (1997) Palaeoenvironmental interpretation of a Triassic-Jurassic boundary section from western Austria based on palaeoecological and geochemical data. Palaeogeography, Palaeoclimatology, Palaeoecology 136: 79-95

McRoberts CA Newton CR Allasinaz A (1995) End-Triassic bivalve extinction: Lombardian Alps, Italy. Historical Biology 9: 297-317

Milner AR (1993) Amphibian-grade Tetrapoda. In Benton MJ (ed) The fossil record 2. Chapman & Hall, London pp 665-679

Milner AR (1994) Late Triassic and Jurassic amphibians: Fossil record and phylogeny. In Fraser NC, Sues H-D (eds) In the shadow of the Dinosaurs. Cambridge University Press, Cambridge pp 5-22

Morbey JS (1975) The palynostratigraphy of the Rhaetian stage, Upper Triassic in the Kendelbachgraben, Austria. Palaeontographica B 152:1-75

Mostler H, Scheuring R, Ulrichs M (1978) Zur Mega-, Mikrofauna und Mikroflora der Kossenen Schichten (alpine Obertrias) von Weissloferbach in Tirol unter besonderer Berucksichtigung der in der suessi- und marshi- Zone auftreitenden Conodonten. Osterreichische Akademie der Wissenschaften Erdwissenschaftliche Kommission Schriftenreihe 4: 141-174

Niklas KJ, Tiffney BH, Knoll AH (1983) Patterns in vascular land plant diversification: A statistical analysis at the species level. Nature 303: 614-616

Ogg JG (2004) The Triassic Period. In Gradstein FM, Ogg JG, Smith AG (eds) A geologic time scale 2004. Cambridge University Press, Cambridge, 271-306

Olsen PE, Galton PM (1984) A review of the reptile and amphibian assemblages from the Stormberg of South Africa, with special emphasis on the footprints and the age of the Stormberg. Paleontologica Africana 25: 87-110

Olsen PE, Sues H-D (1986) Correlation of continental Late Triassic and Early Jurassic sediments, and patterns of the Triassic-Jurassic tetrapod transition. In Padian K (ed) The beginning of the age of dinosaurs. Cambridge University Press, Cambridge, 321-351

Olsen PE, Fowell SJ, Cornet B (1990) The Triassic/Jurassic boundary in continental rocks of eastern North America; a progress report. Geological Society of America Special Paper 247: 585-593

Olsen PE, Shubin NH, Anders MH (1987) New Early Jurassic tetrapod assemblages constrain Triassic-Jurassic tetrapod extinction event. Science 237: 1025-1029

Olsen PE, Smith JB, McDonald NG (1998) Type material of the type species of the classic theropod footprint genera *Eubrontes, Anchisauripus,* and *Grallator* (Early Jurassic, Hartford and Deerfield basins, Connecticut and Massachusetts, U. S. A.). Journal of Vertebrate Paleontology 18: 586-601

Olsen PE, Kent DV, Sues HD, Koeberl C, Huber H, Montanari A, Rainforth EC, Powell SJ, Szajna MJ, Hartline BW (2002a) Ascent of dinosaurs linked to an iridium anomaly at the Triassic-Jurassic boundary. Science 296: 1305-1307

Olsen PE, Koeberl C, Huber H, Montanari A, Fowell SJ, Et-Touhani M, Kent DV (2002b) The continental Triassic-Jurassic boundary in central Pangea: recent progress and preliminary report of an Ir anomaly. Geological Society of America Special Paper 356: 505-522

Orbell G (1973) Palynology of the British Rhaeto-Liassic. Bulletin Geological Society Great Britain 44: 1-44

Orchard MJ (1991) Upper Triassic conodont biochronology and new index species from the Canadian Cordillera. Geological Survey Canada Bulletin 417: 299-335

Orchard MJ (2003) Changes in conodont faunas through the Upper Triassic and implications for boundary definitions. Geological Association of Canada, Vancouver 2003 Meeting, Abstracts Volume: 28

Orchard MJ, Carter ES, Lucas SG, Taylor DG (2007) Rhaetian (Upper Triassic) conodonts and radiolarians from New York Canyon, Nevada, USA. Albertiana 35: 59-65

Pálfy J (2003) Volcanism of the central Atlantic magmatic province as a potential driving force in the end-Triassic mass extinction. AGU Geophysical Monograph 136: 255-267

Pálfy J, Mortensen JK, Carter ES, Smith PL, Friedman RM, Tipper HW (2000) Timing the end-Triassic mass extinction: First on land, then in the sea? Geology 28: 39-42

Pálfy J, Demény A, Haas J, Carter ES, Görög A, Halász D, Oravecz-Scheffer K, Hetényi M, Márton E, Orchard MJ, Ozsvárt P, Vetö I, Zajzon N (2007) Triassic-Jurassic boundary events inferred from integrated stratigraphy of the Csővár section, Hungary: Palaeogeography, Palaeoclimatology, Palaeoecology 244: 11-33

Pedersen KR, Lund JJ (1980) Palynology of the plant-bearing Rhaetian to Hettangian Kap Stewart Formation, Scoresby Sund, East Greenland. Review of Palaeobotany and Palynology 31: 1-69

Racki G (2003) Silica-secreting biota and mass extinctions: survival patterns and processes. Palaeogeography, Palaeoclimatology, Palaeoecology 154: 107-132

Rakús M (1993) Late Triassic and early Jurassic phylloceratids from the Salzkammergut (Northern Calcareous Alps). Jahrbuch Geologische Bundes-Anstalt 136: 933-963

Rigo M, Preto N, Roghi G, Tateo F Mietto P (2007) A rise in the carbonate compensation depth of western Tethys in the Carnian (Late Triassic): Deep-water evidence for the Carnian pluvial event. Palaeogeography, Palaeoclimatology, Palaeoecology 246: 188-205

Roniewicz E, Morycowa E (1989) Triassic Scleractinia and the Triassic/Liassic boundary. Memoir Association Australasian Palaeontologists 8: 347-354

Ruckwied K, Götz AE, Pálfy J, Michalík J (2006) Palynomorph assemblages of Triassic/Jurassic boundary key sections of the NW Tethyan realm: Evidence for climatic change. Volumina Jurassica 4: 297

Sandoval J, O'Dogherty L, Guex J (2001) Evolutionary rates of Jurassic ammonites in relation to sea-level fluctuations. Palaios 16: 311-335

Schäfer P, Fois E (1987) Systematics and evolution of Triassic Bryozoa. Geologica et Palaeontologica 21: 173-225

Schootbrugge B, van de Tremolada F, Rosenthal A, Bailey TR, Feist-Burkhardt S, Brinkhuis H, Pross J, Kent DV, Falkowski PG (2007) End-Triassic calcification crisis and blooms of organic-walled 'disaster species'. Palaeogeography, Palaeoclimatology, Palaeoecology 244: 126-141

Schuurman WML (1977) Aspects of late Triassic palynology; 2, Palynology of the 'Gres et schiste a Avicula contorta' and 'Argiles de Levallois' (Rhaetian) of northeastern France and southern Luxemburg. Review of Palaeobotany and Palynology 23: 159-253

Schuurman WML (1979) Aspects of Late Triassic palynology. 3. Palynology of latest Triassic and earliest Jurassic deposits of the northern limestone Alps in Austria and southern Germany, with special reference to a palynological characterization of the Rhaetian stage in Europe. Review of Palaeobotany and Palynology 27: 53-75

Sepkoski JJ Jr (1982) Mass extinctions in the Phanerozoic oceans: A review. Geological Society of America Special Paper 190: 283-289.

Shubin NH, Olsen PE, Sues H-D (1994) Early Jurassic small tetrapods from the McCoy Brook Formation of Nova Scotia, Canada. In Fraser NC, Sues HD (eds) In the shadow of dinosaurs: Early Mesozoic tetrapods. Cambridge University Press, Cambridge pp 242-250

Signor PW III, Lipps JH (1982) Sampling bias, gradual extinction patterns and catastrophes in the fossil record. Geological Society of America Special Paper 190: 291-296

Sigurdsson H (1990) Assessment of atmospheric impact of volcanic eruptions. Geological Society of America Special Paper 247: 99-110

Simms MJ, Ruffell AH (1989) Synchroneity of climatic change and extinctiosn in the Late Triassic. Geology 17: 265-268

Simms MJ, Ruffell AH (1990) Climatic and biotic change in the Late Triassic. Journal Geological Society, London, 147: 321-327

Skelton PW, Benton MJ (1993) Mollusca: Rostroconchia, Scaphopoda and Bivalvia. In Benton MJ (ed) The fossil record 2. Chapman & Hall, London, 237-263

Stanley GD Jr (1988) The history of early Mesozoic reef communities: a three-step process. Palaios 3: 170-183

Stanley GD Jr (2001) Introduction to reef ecosystems and their evolution. In Stanley GD Jr (ed) The history and sedimentology of ancient reef systems. Kluwer Academic/Plenum Publishers, New York, 1-39

Stanley GD Jr, Beauvais L (1994) Corals from and early Jurassic coral reef in British Columbia—refuge on an oceanic island reef. Lethaia 27: 35-47

Sweet WC (1988) The Conodonta. Clarendon Press, New York, 212 pp

Szajna MJ, Silvestri SM (1996) A new occurrence of the ichnogenus *Brachychirotherium*: Implications for the Triassic-Jurassic mass extinction event. Museum of Northern Arizona Bulletin 60: 275-283

Tanner LH, Lucas SG, Chapman MG (2004) Assessing the record and causes of late triassic extinctions. Earth-Science Reviews 65: 103-139

Taylor DG, Boelling K, Guex J (2000) The Triassic/Jurassic System boundary in the Gabbs Formation, Nevada. In Hall RL, Smith PL (eds) Advances in Jurassic research 2000. Trans Tech Publications LTD, Zurich, 225-236

Taylor DG, Guex J, Rakus M (2001) Hettangian and Sinemurian ammonoid zonation for the western Cordillera of North America. Bulletin de Géologie de l'Université de Lausanne 350: 381-421

Teichert C (1988) Crises in cephalopod evolution. In Marois M (ed) L'évolution dans sa Réalité et ses Diverses Modalités. Fondation Singer-Polignac, Paris pp 7-64

Thulborn T, Turner S (2003) The last dicynodont: An Australian Cretaceous relict. Proceedings Royal Society London B

Tipper HW, Carter ES, Orchard MJ, Tozer ET (1994) The Triassic-Jurassic (T-J) boundary in Queen Charlotte Islands, British Columbia defined by ammonites, conodonts, and radiolarians. Geobios Mémoire Special 17: 485-492

Tomašových A, Siblík M (2007) Evaluating compositional turnover of brachiopod communities during the end-Triassic mass extinction (Northern Calcareous Alps): Removal of dominant groups, recovery and community reassembly. Palaeogeography, Palaeoclimatology, Palaeoecology 244: 170-200

Tozer ET (1981) Triassic Ammonoidea: Classification, evolution and relationship with Permian and Jurassic forms. In House MJ, Senior JR (eds) The Ammonoidea: Systematics Association Special Volume 18: 65-100

Traverse A (1988) Plant evolution dances to a different beat. Historical Biology 1: 277-301

Ulrichs M (1972) Ostracoden aus den Kössener Schichten und ihre Abhängigkeit von der Ökologie. Mitteilungen der Gesellschaft der Geologie- und Bergbaustudenten in Österrich 21: 661-710

Vermeij GJ (1977) The Mesozoic marine revolution: Evidence from snails, predators and grazers. Paleobiology 3: 245-258

Vermeij GJ (1983) Evolution and escalation: An ecological history of life. Princeton University Press, 527 pp

Vishnevskaya V (1997) Development of Palaeozoic-Mesozoic Radiolaria in the northwestern Pacific rim. Marine Micropalaeontology 30: 79-95

Visscher H, Van Houte M, Brugman WA, Poort RJ (1994) Rejection of a Carnian (late Triassic) "pluvial event" in Europe. Review of Palaeobotany and Palynology 83: 217-226

Ward PD, Haggart JW, Carter ES, Wilbur D, Tipper HW, Evans T (2001) Sudden productivity collapse associated with the Triassic-Jurassic boundary mass extinction. Science 292: 1148-1151

Ward PD, Garrison GH, Haggart JW, Kring DA, Beattie MJ (2004) Isotopic evidence bearing on Late Triassic extinction events, Queen Charlotte Islands, British Columbia, and implications for the duration and cause of the Triassic-Jurassic mass extinction: Earth and Planetary Science Letters 224: 589-600

Ward PL, Botha J, Buick R, De Kock MO, Erwin DH, Garrison GH, Kirschvink JL, Smith R (2005) Abrupt and gradual extinction among Late Permian land vertebrates in the Karoo basin, South Africa. Science 307: 709-714

Warrington G (2005) The Triassic-Jurassic boundary-a review. In L'Hettangian à Hettange de la science au patrimoine. Réserve Naturelle Hettange-Grande, Moselle, pp.11-14

Warrington G, Cope JCW, Ivimey-Cook HC (1994) St. Audrie's Bay, Somerset, England: A candidate global stratotype section and point for the base of the Jurassic System. Geological Magazine 131: 191-200

Weems RE (1992) The "terminal Triassic catastrophic extinction event" in perspective: A review of Carboniferous through Early Jurassic terrestrial vertebrate extinction patterns. Palaeogeography, Palaeoclimatology, Palaeoecology 94: 1-29

Whiteside JH, Olsen PE, Kent DV, Fowell SJ, Et-Touhami M (2007) Synchrony between the Central Atlantic magmatic province and the Triassic-Jurassic mass-extinction event? Palaeogeography, Palaeoclimatology, Palaeoecology 244: 345-367

Williford KH, Ward PD, Garrison GH, Buick R (2007) An extended organic carbon-isotope record across the Triassic-Jurassic boundary in the Queen Charlotte Islands, British Columbia, Canada. Palaeogeography, Palaeoclimatology, Palaeoecology 244: 290-296

9 Cenomanian/Turonian mass extinction of macroinvertebrates in the context of Paleoecology; A case study from North Wadi Qena, Eastern Desert, Egypt

Ahmed Awad Abdelhady

Geology Department, Faculty of Science, Minia University, Minia, Egypt, alhady2003@yahoo.com

9.1 Introduction

The paleoenvironment of north Wadi Qena area represents an obvious importance in assessing the effects on the Late Cretaceous biota; such a paleoenvironmental interpretation would include biotic evidence from invertebrate fossils and from geologic or sedimentary features such as paleosols and sequence stratigraphy interpretations. Numerous authors have demonstrated that aspects of modern macrofaunas are highly correlated with environmental variables, suggesting that invertebrate fossils are useful as paleoenvironmental indicators. Major extinction events are cyclic and may be forced by a similar, if not identical, mechanism raised the intriguing possibility that mass extinctions could be readily understood. However, investigations of various Phanerozoic mass extinctions (e.g., Kauffman 1988) have shown that comparisons are not as simple as anticipated. These studies have documented substantial differences between many of the mass extinctions in terms of their biotic, lithologic and geochemical fabrics, despite these differences, the broad pattern of repopulation appears to be comparable between biotic crises, and certain events have a very correlative biotic response from a wide variety of perspectives. This study aims to put constraints on environmental changes that affect the biota based on macroinvertebrates.

The mass extinction in macroinvertebrates across the Cenomanian/Turonian transition took place during a peak global greenhouse interval, attributed to maximum sea level rising (the most intense Phanerozoic flooding event). During the C/T interval, the see was probably 255 m higher (Haq et al. 1987) and shelf areas were twice as large as those today. Abrupt environmental changes include the absence of polar ice caps during the peak of a Greenhouse cycle. Atmospheric CO_2 at least four times preset levels; global warm, expansion of the oceanic oxygen minimum zone (OAE) to the deep ocean floor and epicontinental sea habitats, initiating trace element advection and chemical stirring of the oceans and oceanic impacts of meteorites and/or comets as part of the Cenomanian impact (Olsson et al., 2001). Herein the boundary represents a great mass extinction event. High global sea level during the Mid and Late Cretaceous are related to fast Atlantic sea-floor spreading (Thurow et al. 1992, p. 269);

Although many Egyptian authors studied the C/T transition (e. g. Kora and Hamama 1987; Luger and Groscke 1989; Kassab 1999; Abdelhamid and El-Qot 2001; Galal et al. 2001; El-Hedeny and Nafee 2001; Zakhera 2002a, 2002b, 2003; Abdelhamid and Azab 2003; Khalil and Meshaly 2004; Abdel-Gawad et al. 2004), mass extinction at the C/T transition never been analyzed. The main targets of the present work are to verify environmental, faunal, and architectural changes during phases of growth of the late Cenomanian/Turonian platform north of Wadi Qena, Eastern Desert, Egypt to investigate the interaction with eustatic sea-level, oxygenation, environmental energy and temperature changes and to figure out other possible controlling factors in the mass extinction process.

Cenomanian/Turonian mass extinction received extensive investigations backs to the original surveys of Harries (1993, 1999); Harries and Kauffman (1990); Harries and Crispin (1999); Kauffman (1988, 1995); Kauffman and Harries (1996). Detailed investigations of macroinvertebrate fossil groups are needed to highlight the impact of the different faunal groups to detect faunal differences between Cenomanian/Turonian assemblages and to reconstruct in which way these assemblages are influenced by platform flooding and crisis. Some benthic macrofossil groups constitute sensible indicators for environmental changes, such as variations in water depth, oxygen content, nutrient supply or water energy and various studies about these groups as ecological proxies were done. A focus lies on the questions, why many larger benthic fauna disappeared during late Cenomanian times and did not resettle on Turonian Moreover; assemblages of oyster bivalve constitute environmental proxies, do their occurrence and distribution indicate changes of water depth, water temperature and nutrient supply?

9.2 Geographical and geologic setting

The study area is located at the upper reaches of Wadi Qena and Wadi Tarfa, between latitudes 28° 00" and 28° 25" N, bounded to the north by the road of El Sheikh Fadl-Ras Gharib, and longitudes 32° 10" and 32° 30" E; bordered by the Basement complex to the East (Fig. 1). It lies at the Northern rim of the Wadi Qena, Eastern Desert, Egypt. The Upper Cretaceous strata overlain the Malha Formation of Early Cretaceous (Aptian ?), The rock units expressed here are: El-Galala Group (Albian-early Turonian), and it's divided into; Raha Formation (Latemost Albian-Middle Cenomanian) and Abu Qada Formation (Late Cenomanian-early Turonian) and Uum Omieyid (Middle-Late Turonian. The Upper Cretaceous strata form small hills, and there thickness increases in a southward direction with dip up to 15° (Fig. 1).

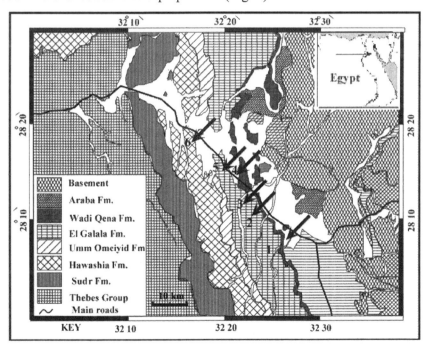

Fig. 1. Simplified geologic map of the study area showing the location of the measured sections (Simplified after Klitzsch and Hermmina (1989))

During the Late Cretaceous, north Wadi Qena area was a part of the broad northern shelf of the Afro-Arabian Plate. The Gulf of Suez and Gulf of Aqaba rifts, which bound the Sinai microplate today, were still closed. While transgressive movements from the late Turonian onwards affected north Eastern Desert, northern Sinai and the adjacent Negev, North Wadi Qena, southern and central Sinai are believed to have remained tectonically rather quiet through out the Mesozoic and Early Tertiary. While north Sinai are part of the `Syrian Arc' (Krenkel 1924) which represents an intraplate fold belt extending from Egypt to Syria and Jordan, and was formed by Late Cretaceous to Recent inversion of Late Triassic half-grabens (Shahar 1994). North Wadi Qena area lies on the southern tectonically inactive block and is termed `stable shelf, the sections studied here are situated on the stable shelf and represent the most inner shelf area which subjected to many landward exposure during Late Cretaceous.

9.3 Material and methods

Six stratigraphic sections were chosen on the study area according to their richness on macroinvertebrate fauna, samples from all sections were collected by the same team using the same methods, one hundred and eighty species were identified including seventy three bivalves, forty-three gastropods, twenty-six ammonites and thirty-eight echinoids representing both local and worldwide taxa. All collected groups were prepared, identified and special emphasis was given to the vertical and lateral distribution and ecological preferences of this identified taxa. The fossils and rock specimen's completely discussed and figured in details in the master thesis of the author and stored in the collection of the Paleontology museum, Geology Department, Faculty of science, Minia University, Minia, Egypt, coded from AA1 to AA180.

The approach to data collection was to apply high-resolution stratigraphical data collection techniques (sensu Kauffman 1988). This approach is critical to mass extinction studies inasmuch as it allows temporally rapid changes in biotic and paleoenvironmental parameters to be recognized. The locations of the sections analyzed are shown in Figure 1. It should be noted that the data come from a number of sections, and although they are not global in distribution, they represent a much more robust dataset than simply a single section, where local changes cannot be are somewhat more expansive because of the greater number of sections and depositional environments they encompass (Harries and Crispin 1999). For both mass extinction events and the paleoecology, data are based on two primary sources: stratigraphic range and abundance data (Figs. 13-17 and Tabs. 2-5). In both cases, the data grouped using a combination of taxonomy and ecological preferences.

The taxa have been further subdivided, following the approach initially employed, by Harries and Kauffman (1990), into various species 'modes'; Cenomanian species (Extinct), Cenomanian-Turonian species (Survivors) and Turonian (Repopulated/ Originated). These stratigraphic data tallied to obtain diversity levels for the various mass extinction intervals as well as abundance and species richness plots for ecologic index taxa, which depict the short-term biotic changes. These biotic crises consist of three main components: the extinction, the survival, and the recovery intervals. They are differentiating based on the interplay between extinction (E) and origination (O) rates (Keller et al. 2002). In the extinction interval we found E>O and the diversity declines dramatically. In the survival interval, E=O and values for both are relatively low in combination with diminished diversity and abundance. As the survival interval progresses, O progressively increases relative to E. As the name implies, this interval is primarily dominated by surviving taxa and by an increasing number of newly evolved species from surviving lineages (sensu Raup 1984). The recovery interval is dominated by origination events, therefore O>E (Harries and Crispin 1999).

9.4 Correlation of diversity and preservation

Before entering into the details of faunal abundance and diversities, a few significant characteristics of Late Cretaceous macroinvertebrates need to mention. Macroinvertebrates are highly susceptible to selective dissolution, both before and after burial in the sediment but especially during late digenesis and deep burial. (The extent of preservation of macroinvertebrates assemblages therefore is strongly dependent on the digenetic microenvironment.

In general terms, well-preserved samples are highly diverse (easily over 100 morphotypes) and poorly preserved samples have a low diversity (10 morphotypes less). This demonstrates that the absence of a morphotype in part of a column does not necessarily have biochronologic significance and it seems evident that high-diversity species are less solution-resistant than low- 4 diversity species. It can observe generally that common or abundant morphotypes are longer ranging than rare ones. Low-diversity species often are common to abundant; high-diversity species are often rare (there are exceptions). At our present stage of knowledge, these correlations seem mainly preservation-controlled. An ecologic signal (bathymetric and/or latitudinal distribution) nevertheless may be hidden in these relationships. Its solution has to await a well- established biochronology, which allows a correlation of samples from different paleogeographic (paleobathymetric) positions.

9.5 Biostratigraphy

Biostratigraphic subdivision of the interval considered here was based mainly on, Ammonite, Bivalvia and, to a lesser extent Gastropoda and Echinoidea. Due to the differential thickness of Upper Cretaceous strata in the studied sections the results are presented as a composite biochronological scheme that represent a comprehensive interpretation of the macrofaunal data in a regional sense. Several ammonite-bearing horizons occur within the uppermost Albian-Santonian succession of the study area, which can easily compare with occurrences of adjacent areas (Israel and Jordan). We follow the Ammonite zonation scheme of southern Europe (Hardenbol et al. 1998). Extensive occurrence of Bivalvia started from latemost Albian to the Santonian. Gastropoda have little importance on biostratigraphical zonation as they almost of long ranging, their vertical distribution takes with care with regarding to associated ammonites or oysters. Echinoid is the same as gastropod and be of low significant value in biostratigraphical zonation, their importance stand where their occurrences correlates with those of the adjacent areas (Table 1, Abdelhady in press).

Table 1. Macroinvertebrate biozonal scheme for the Upper Cretaceous of North Wadi Qena strata compared to standard of Hardenbol et al. 1998

Stage	Bivalvia	Ammonites	Standard
Santonian	Oscilopha dichotoma	Texanites texanus	
Coniacian	Meretrix plana	Metatissotia fourneli	
Turonian	Astarte (Triodonta) tenuicostata Plicatula reynesi Neithea dutugei	Coilopoceras requinianaum Choffaticeras Segne Mamite nodosoides P. flexosum	omatissmum kalleisi Turoniense woollgari nodosoides coloradoense
Cenomanian	Pycnodonte (Ph.) vesicularis Exogyra (C.) olisiponensis Ceratostreon flabellatum Rhynchostreon suborbiculatum	V. cauvini M. geslinianum N. vibrayeanus	juddii geslinianum guerangeri rhotomagensi
Albian	Nucula (N.) Margaritifera	Stoliczkaia sp.	

9.6 Results

9.6.1 Ecological preferences

Paleoecology of the Late Cretaceous macrofauna is formally established. The macrofaunal categories employed to infer paleoenvironmental conditions. The methods of establishment of the optimum conditions for the paleoecology is more precisely approaches by analyzed the association grouply. Substantial efforts of the past decade have resulted in a large database of macroinvertebrate ecology and several attempts of interpretations (i.e. Hallam 1967; Néreaudeau 1995; Huber, 1995; Néraudeau et al. 1997; Dhondt et. al 1999; Seeling and Bengston 1999; Walker 2001; Videt and Platel 2003).

Bivalvia

In the Cenomanian/Turonian strata of north Wadi Qena area, Bivalvia occasionally oysters occur in accumulations, which generally represent shallow-marine environments. With the beginning of Turonian age, oysters become scarce or vanish, this may reelect the fact that, following the Cenomanian/Turonian transgression, conditions became too deep for oysters to flourish, in addition the same criteria assign for the Turonian oysters of North Africa (Dhondt et al. 1999) and Brazil (Seeling and Bengston 1999). The attachment area in most of the investigated species is very small, or even absent, as in *Rhynchostreon mermeti* indicate partly buried in, soft substrates without being attached. Therefore, there are indicating that at least parts of oysters represent a transported fauna. Size differentiation of the individuals has observed at different sections. Thus, in section 1 specimen are generally smaller than in the other sections. Size differentiation is possibly the result of sorting by sedimentary processes. According to (Mancini 1978), a transported micromorph fauna formed when the smaller individuals of an assemblage are winnowed out and concentrate separately.

Oysters are strongly inequivalve. The lower, left valve is heavy, cup-shaped and more durable than the upper, right valve, which is thinner, lighter and more fragile. An oyster community buried in situ, or transported only a short distance, would consist of approximately equal numbers of left and right valves. Transport would be expected to destroy or remove the lighter, more fragile right valves and deposit the heavier left valves at a shorter distance from the original site.

envets / species	Salinity		Oxygen			Turbidity			Substrate										Indur-tion			Agitation		
	euryhaline (18-40 PSU)	stenohaline (35 PSU)	normal	modrate	anoxic	strong	modrate	fable	variable agency	biogenic	sand	pebble	gravel	course sand	fine sand	silt	marl	argiles	soft	fast	modrate	high	modrate	low
Pycnodonte (P.) vesicularis		●	●			●	●	●	●			●	●	●	●	●	●			●	●	●	●	
Exogyra conica		●	●	●	●	●	●	●	●	?	●	?	•	•	●	●	●	•		●	●		●	●
Exogyra ribosa		●	●	●			●			?	?	?	?	?	●	●	●		●	●	●		●	●
E. (C.) olisiponensis		●	●				●	●	●	?	●	●	●	●	●	●	●		?	●	●		●	
Amphidonte obliquatum		●	●				●	●	●	●	●	●	●	●	●	●	●		●	●	●		●	●
Ceratostreon flabellatum		●	●				●	●	●	●	●	●	●	●	●	●	●		●	●	●		●	●
Ceratostreon texanum	?	●	●	●	●	●	●	●	•	?	•	•	●	●	●	●	●			●	●	●	●	●
Hyotissa semiplana	?	●	●	●	●	●	●	●	●	●	●	●	●	●	●	●	●	●	?	●	●	●	●	●
Ilymatogyra arietina		●	●	●	?	●	●	•	●	?	●	●	●	●	●	●	●	•	●	●	●		•	●
Ilymatogyra (A.) african	?	●	●	●	*?		●	●	?	●	?	?	●	●	●	?	?			●	●	●	●	●
Rhynchostreon mermeti		●	●	●	●		●	●		?	?	?	?	?	?	●	●		●	●	?		●	●
R . suborbiculatum		●	●	●	●		●	●		?	?	?	?	?	?	●	●		●	●	?		●	●
Gyrostrea delettrei	●	●	●	●	*	●	●	●	●	●	●	●	●	●	●	●	●	●		●	●		●	●
Lopha sp.	•	●	●	●		●	●	●	●	●	●	●	●	●	●	●	●	●		●	●	●	●	●
Oscilopha acanthoantha		●	●	●	●		●	●		●	●	●	●	●	●	●	●		●	●	●	●	●	●
Oscilopha dichotom		●	●	●	•		●	●	●		●	●	●	●	●	●	●		●	●	●	●	●	●
Agerostrea ungulata		●	●	●			●	●		?	?	?	?	●	●	●	●			●	●	●	●	●
Rastellum (A.) carinatum	?	●	●	●	*?		●	●	?	●	?	?	●	●	●	?	?			●	●	●	●	●
Curvostrea isidis		●	●	●	●		●	●		?	?	?	?	?	?	●	●		●	●	?		●	●
Curvostrea heinzi	●	●	●	●	*	●	●	●	●	●	●	●	●	●	●	●	●	●		●	●		●	●
Curvostrea rouvillei		●	●	●			●	●		?	?	?	?	?	●	●	●			●	●	●	●	●

Fig. 2. Ecological parameters for some oyster species, combined from literatures

In the studied material, right and left valves are rarely found together. The much smaller number of right valves in the assemblage is attributed to the fragility and breakage, is a reliable means of determining the relative amount of transport an oyster or other bivalve has undergone since death. In particular, smaller specimens of *Pycnodonte (Phygraea) vesiculosa, Rhynchostreon mermeti* and *Ilymatogyra (Afrogyra) africana* are not found in life position and are mostly disarticulated. They are only partly fragmented. In contrast, nearly all larger specimens, mainly of *Exogyra (Costagyra) olisiponensis* and *I. (A.) africana,* are preserved unbroken and articulated. This indicates transport under high-energy conditions, but of short duration, as abrasion and breakage is usually low and lower weight as described above. The degree of disarticulation, along with abrasion and breakage, is a reliable means of determining. Definite ecological parameters can be determined based on index oyster species (Figs. 3, 4).

Different ecological conditions characterize the *Neithea*. Herein *Neithea* was found mainly in carbonate rocks representing Circalittoral environments at the early Turonian. All species are recliners that rested on soft substrates or with their convex right valve partly buried, without being attached. Some of the *Neithea* occurrences in north Wadi Qena appear to represent un-transported faunas or transported only a short distance as they consist of approximately equal numbers of left and right valves. The degree of disarticulation, along with abrasion and breakage, is a reliable indication of the degree of transport a bivalve has undergone since its death.

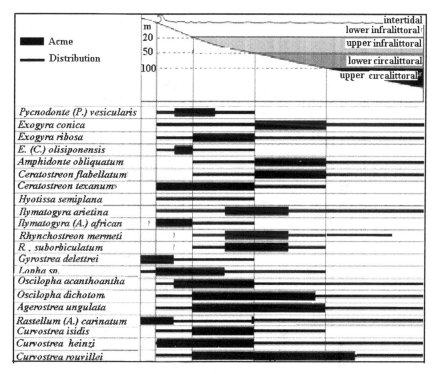

Fig. 3. Distribution of oyster species within the depositional environments

Gastropoda

Gastropod represent well diverse and dense macrofaunal group. *Turritillia* are dominating at early Cenomanian as internal molds, while Tylostoma are dominate by the beginning of Turonian. The *turritella* are a good indicator for cold water while the Tylostoma are of high temperature gradients, this reflect the fact that the Cenomanian/Turonian boundary marked by warming conditions. The gastropod fauna represented by irregular densities from the beginning of the base of the measured sections to the top during all intervals from latest Albian to the late Santonian, important ecological interpretations for index gastropod genera shown in Table 2.

Table 2. Ecological parameters for gastropod genera. CP predatory carnivores; CB browsing carnivores; HO herbivores omnivores; HP plant & algae herbivores; SU suspension feeders; D deposited feeders; IS shallow infaunal ; In infuanal; E epifaunal, * analogic, compiled from literatures

Genus	Mode of Life	Trophic style	Environment		
			Littoral	Sublittoral	Outer
				Infra-	Circa-
Turritella	IS	SU			
Mesalia	In	SU			
Cerithiella	E*	CB			
Cimolithium	E*	HP			
Cerithium	E	HM/HR			
Turriscala	In	CB*			
Nerinea	IS	HP*			
Aptyxiella	IS	HP*			
Aporrhais	E	D			
Columbellina	E*	HO*			
Pterodonta	E*	CP			
Strombus	SU	HO			
Rhynchocypraea	IS	CP			
Gyrodes	E*	CP*			
Tylostoma	E	CP*			
Fasciolaria	E	CP			
Acteonella	E	CP			

Ammonites

No ammonites defined before the late Cenomanian except the *Stoliczkaia* sp., which represented by small fragments occurred at the lower part of Raha Formation, This suggests a very shallow marine deposition for early-middle Cenomanian. Shell fragments of ammonites collected from the study area, occasionally at late Albian and at the Coniacian-Santonian, are related to agitated wave conditions and characterize infralittoral environment at episode of fast subsidence. The consistent occurrence of ammonites regarded at the uppermost Cenomanian (where oysters go vanish) indicates deeper marine condition against late Cenomanian background. Material collected here represent both complete and incomplete internal mold but these mold with clear ribs, nodes sculptures and clear sutures. In general, the proportion of vascoceratides ammonites is known to increase with depth and, according to several authors, an oceanic deepwater character would therefore be indicated for Cenomanian/Turonian boundary, deposition on the middle or lower slope therefore assumed for the late Cenomanian at this site. The poor foraminiferal fauna presumed to reflect Oceanic Anoxic Event OAE. This agrees with the correlation of biostratigraphical events between the measured sections.

Echinoidea

Extensive occurrence for this group started with lower boundary of Raha Formation. *Hemiaster* dominated from the bigining but the oldest one is somewhat badly preserved and well withered and hence they obscure and scratched in almost cases, whoever in progress many individuals preserved well. Specimen test affected by the type of sediments buried in, the yellowish ones found in sandy substrata, the gray shale give the gray tone to the imbedded shells. Dark shells collected from organic rich beds while the light echinoid test found in the carbonate beds. Late Turonian contain unusual very small accumulation of *Hemiaster Herbert*, which indicate anoxic or general unsuitable conditions (Néraudeau et al. 1997).

9.6.2 Environmental analysis

Raha Formation

This formation comprises the following three members:

(1) Abu Had Member (3.50 meters); which made up by about glauconitic sandstones, laminated gypsiferous mud and variegated clay stones with association include oysterid and nuculid Bivalvia, tylostomid Gastropoda, hemiasterid Echinoidea and *stoliczkaia* sp. Fragments. All fauna characterized by small size. The age of Abu Had Member is latemost Albian to early Cenomanian age and it is deposited under infralittoral environment with moderate oxygenation percentage, agitated and stenohaline water column;

(2) Mellaha sand Member; comprises the interval of early/middle Cenomanian and built up of sandstone intercalated with mudstone and ferruginous sandstone beds. Hemiasterid Echinoidea, Turritellid and cerithillid Gastropoda with consistent occurrence of *Rhynchostreon suborbiculatum* which indicate lower infralittoral environment with restricted circulation, normal oxygenation level, euryhaline water column and high energy conditions with comparison to the Abu Had Member;

(3) Ekma Member; represented by sand biostatics, shallow water shales and sandy shale rich on turritellid Gastropoda of cold water and oyster bank of *Ceratostreon flabellatum* suggest deposition under agitated infralittoral with normal oxygen content and stenosaline water column.

Abu Qada Formation

In general, more calm late infralittoral to late circalittoral environments than those of Raha Formation, where individuals more developed. Oyster accumulations suggest bars on shoal environment while bioturbation and flat bedding reflects subtidal facies. The faunal groups attain thus gigantic forms. Consistent occurrence of vascoceratid ammonites, tylostomatiid gastropodas and regular echinoids indicates open marine conditions with open circalattoral (a worldwide transgressive phase spanning the early Turonian, Haq et al. 1987). Sudden disappearance of oysters encourages this idea. Low faunal diversity and mass extinction of this horizon related to the OAE of late Cenomanian-Turonian Boundary (Fig. 4). Abrupt increase in temperature represented by organic matter rich shales of the Abu Qada Formation.

Umm Omieyid Formation

Fine-grain sands, trough cross-stratified sandstone and fissile shale indicate deposition in slightly agitated conditions reveled with coursing up sandstone beds with shell fragments of ammonites despite of their thickening walls. The ammonites fragmentation was affected by agitation at the shoreface where deposition took place; the further regression with high agitated condition at this time is the favorable causes of above mentioned observations. Appearance of Hemiasterid echinoid suggested a phase similar to that of early Cenomanian. *Hemiaster herberti* appear with big bank in small size (not more than 12mm length) indicate Sharpe decrease in nutrients material and general toxic conditions.

Hawashiya Formation

Hawashiya Formation is characterized by a decrease in faunal diversity and high reflux of course clastics, which indicate nearshore environment of high energy agitated conditions. Appearance of gypsiferous lamina indicates lagonal environment and arid conditions. Appearance of oyster bivalve again suggested stability in environment and homogeneous oxygenation percentage. They nevertheless could be met in the early circalittoral. Their accumulation seems rather localized in the early infralittoral. The environments where they generally could be met are silt or marls substratum or by the indurations of the substratum, or by the extensive turbidity, or again by eventual problems of anoxic conditions (Fig. 3).

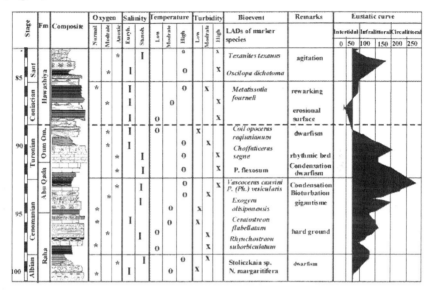

Fig. 4. Environmental parameters and eustatic sea curve, dash line represent erosional surface, (P=*Pseudoaspidoceras*, T= *Texanites*; V=*Vascocerass*)

9.6.3 Mass extinction analysis

Species extinctions

One hundred and eighty species with high individual density examined. Most species disappear at the C/T boundary, A total of 6.6% of Cenomanian species range well into the early Turonian and are considered as survivors as discussed below. Thus, we consider 76.4% of species as extinct at or near the C/T boundary (*Vascocears cauvini/ Pseudoaspidoceras flexosum* Zone). The combined relative abundance of all ammonites genera average less than 50% of the total Turonian assemblage (Tables 3, 4). The great species extinction and relative abundance pattern was observed in other macrofaunal groups, 94.9% of Bivalvia and 75% of Gastropoda extinct at the C/T and 63.4% of Echinoidea died out before the beginning of the early Turonian (Fig. 5). Nearly all of the extinct species have large morphologies, highly ornamented tests, and their geographic distributions are restricted to low and middle latitudes. We consider these taxa as ecological specialists well adapted to tropical and subtropical environments, but intolerant of environmental changes, including fluctuations in temperature, nutrients, oxygen and salinity as previously discussed. The number of extinct species 'E' below the C/T boundary larger than repopulated/originated 'O', E>O.

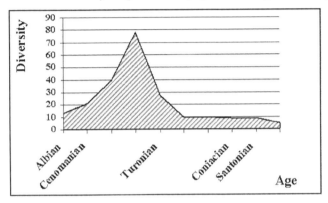

Fig. 5. Total faunal diversity of macroinvertebrates represents the demise at the C/T

The subdivision of extinction intensity into taxonomic groups shows a fairly devastating impact of the mass extinction at least upon the biota of north Wadi Qena (Figs. 6, 7). Extinction intensity among *Vascoceras, Turritella, Tylostoma* and *Hemiaster* are strongly different. It is being 100% at the specific level of *Turritella* and oysters; 80% of *Hemiaster*. *Tylostoma* and *Vascoceras* in the other hand their diversity and flourishing took place at the C/T transition and after where percent of Extinct is limited to 25% for *Tylostoma* and 31.5% for *Vascoceras*. Theses suggest ecological preferences for new ecological changes (high temperatures, deepening water column) and ability to tolerate anoxia and salinity. Although *Hemiaster* continue by 20% of its total diversity in the post C/T transition, its ability to tolerate the harm conditions less than other co-occurred groups as they go unusual size "dwarfed fauna" and vanishing by the end of early Turonian where replaced by regular genera including, *Rachiosoma* and *Phymosoma*. Oyster bivalve was the major group affecting by the C/T crises 'highly sensitive', their preferences occurrence restricted to shallow marine conditions.

An examination of the species richness and abundance data shows a substantial change in the distribution of taxa that correspond to the main extinction horizon (Figs. 6, 7 & Tables 5, 6). Prior to the extinction, the diversity distribution between the different taxonomic groups is fairly even, whereas following the extinction the dominance shifts to predominantly epifaunal bivalves and ammonites. This suggests that the C/T event resulted in substantially different ecologic structure as well. The C/T mass extinction resulted in the extinction of 76.4% of the macroinvertebrate (Fig. 5).

Table 3. Diversities of macrofaunal groups within the Albian-Santonian time interval

	Albian	Cenomanian			Turonian			Coniacian	Santonian	
	Late	Early	Middle	Late	Early	Middle	Late	Late	Early	Late
Ammonites	1	0	0	10	9	4	3	1	0	1
Bivalvia	6	4	8	20	3	2	3	2	3	0
Gastropoda	3	9	7	9	2	1	2	2	6	3
Echinoidea	3	3	5	6	7	1	4	6	0	1

Table 4. Total Diversity and Percentages of Extinct, Survivors and Repopulation groups

	Diversity				Percentage			
	Extinct	Survivors	Repop.	Total	Extinct	Survivors	Repop.	Total
Ammonites	8	6	9	23	34.7%	26.1%	39.2%	100%
Bivalvia	73	0	4	77	94.9%	0%	5.1%	100%
Gastropoda	30	4	6	40	75%	10%	15%	100%
Echinoidea	19	1	10	30	63.4%	3.3%	33.3%	100%
Total	130	11	29	170	76.4%	6.6%	17%	100%

Species-level extinction value of C/T is higher for all macroinvertebrate groups. The fabric of the extinction is similar to that of the K/T; subsequent extinction steps sequentially reduce diversity to its nadir at the C/T boundary, abundance trends are similar as well, with initially high diversity and abundances followed by a punctuated reduction throughout the extinction interval. These differences probably reflect:

(1) The increased extinction intensity within epicontinental or more regional settings;

(2) The relative paucity of extinction-resistant taxa within the basin; and

(3) The inclusion of groups that tend to evolve relatively rapidly, at least at the species level.

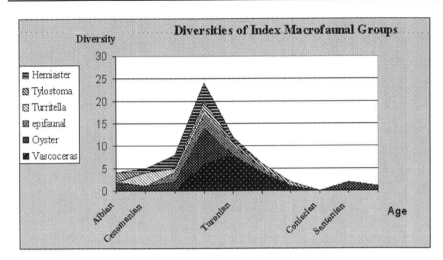

Fig. 6. Dramatic change of diversities of selected index macrofaunal groups at the C/T boundary

Table 5. Diversity and species richness of index macrofaunal groups during the Albian-Santonian interval

		Albian	Cenomanian			Turonian			Coniacian	Santonian	
Vascoceras	Diversity	0	0	0	6	8	4	1	0	0	0
	Richness	0	0	0	58	42	18	20	0	0	0
Oyster	Diversity	2	1	2	8	0	0	0	0	2	1
	Richness	11	255	20	742	0	0	0	0	54	30
epifaunal	Diversity	0	0	2	3	2	1	0	0	0	0
	Richness	0	0	22	45	75	165	0	0	0	0
Turritella	Diversity	0	3	1	1	0	0	0	0	0	0
	Richness	0	62	23	11	0	0	0	0	0	0
Tylostoma	Diversity	2	0	0	2	1	1	1	0	0	0
	Richness	6	0	0	9	8	18	49	0	0	0
Hemiaster	Diversity	0	1	3	4	1	0	0	0	0	0
	Richness	0	18	86	64	211	0	0	0	0	0

Survivors

Only four groups with species-level survivors include: Ammonites (Occasionally Vascoceratriides); epifaunal Bivalvia (*Neithea*, *Inoceramus*, *Plicatula*), and globular Gastropoda (*Tylostoma*); irregular Echinoidea (*Hemiaster* ' if we consider *Hemiaster herberti* as a survivor'), percent survivors among these groups is 42.2%, 33.3%, 25% and 20% respectively (Table 5). Obviously, the C/T event had a substantial impact on the fauna, although the low number of new species derived from surviving lineages, either there were an invasion of immigrants from surrounding areas, especially the Tethyan realm.

Table 6. Total Diversity and Percentages of Extinct, Survivors and Repopulation of index macrofaunal groups

	Diversity				Percentage			
	Extinct	Survivors	Repop.	Total	Extinct	Survivors	Repop.	Total
Vascoceras	6	8	5	19	31.5%	42.2%	26.3%	100%
Oyster	8	0	0	8	100%	0%	0%	100%
epifaunal	3	2	1	6	50%	33.3%	16.7%	100%
Turritella	1	0	0	1	100%	0%	0%	100%
Tylostoma	1	1	2	4	25%	25%	50%	100%
Hemiaster	4	0	1	5	80%	0%	20%	100%

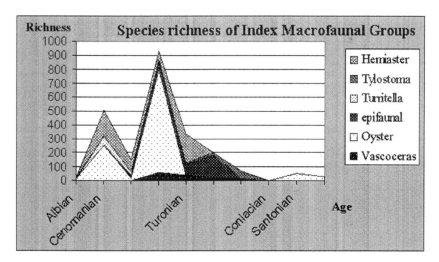

Fig. 7. Species richness of index macrofaunal groups during the Albian-Santonian interval

A smaller number of Cenomanian species range well into the early Turonian at almost sections. The fact that the mass extinction eliminated only ecological specialists having relatively narrow ecological habitats points to a selective mass extinction pattern. The groups, which range into the early Turonian are generally common to abundant in the late Cenomanian. These species are considered Cenomanian survivors because they have been observed to be consistently present in early Turonian sediments of local sections and worldwide ones. Among these *Pseudoaspidoceras pseudonodosoides, Selarilla (S.) juxi, Tylostoma athlyticum, Nerinea Sinaiensis, Tornatella brevicula* and *Tetragramma variolare*.

Morphologically, these taxa are generally small with little or no surface ornamentation. They are geographically widespread and for the most part common to abundant. We consider them ecological generalists, able to tolerate fluctuations in temperature, nutrients, oxygen and Salinity. It is noteworthy that in the examination of the size fraction; the larger the organism size the intense the mass extinction, the number of extinct equal to the originated E=O at the C/T boundary (Fig. 9).

Repopulation

The populations of most Cenomanian survivors dramatically decline in the early Turonian and never recover. Some Cenomanian species, however, thrive after the mass extinction of tropical and subtropical species and the decline of the ecological generalist survivors. Because of the absence of ecological competition as a result of the mass extinction and decline of survivor species, and prior to the establishment of the newly evolving Turonian assemblages the faunal assemblages during the early Turonian are generally present in very low abundances.

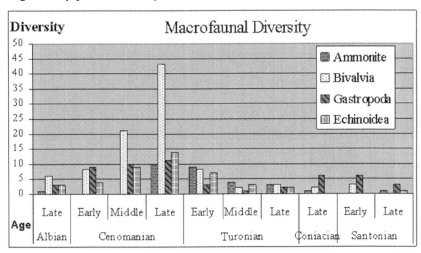

Fig. 8. Relative diversities of macroinvertebrate categories at Albian-Santonian

Cenomanian faunal assemblages of normal shallow neritic near-shore environments differ completely from the Turonian open marine conditions and more variable temperature, oxygen, salinity and nutrient conditions. However, open marine environmental conditions reach a crisis level and reduce normal population diversity they produce opportunistic blooms (e.g., Ammonites) resemble the blooms in the Certaceous/Tertiary boundary (Keller et al. 2002). The newly Turonian genera includes

Vascoceras durandi, Vascoceras obessum, Vascoceras proprium, Pseudoaspidoceras flexosum, Mamites nodosoides, Neoptychites xetriformis, Thomosites rolandii, Coenholectypus turonensis, Rachiosoma irregulare, Phymosoma bbatei, Hemiaster herbeti, Tylostoma gadensis, Tylostoma globosum, Tylostoma cossoni, Tylostoma pallayi, Apyyxalia (Neronoides) subequalus Septifer samiri, Neithea hispanica, Neithea dutrugei, Plicatula batnensis.

The number of extinct taxa for the early Turonian interval less than repopulated E<O. Examination of index genera of Cenomanian oyster, occasionally their species richness (Fig. 10) indicate that, although at every boundary abrupt change occurs in the specific level, the newly abundant species flourish again fastly. This evaluate that changing in abundances occurs simultaneously but return fastly to its normal state but at the C/T boundary environmental changes was great representing platform crises which and inhibit normal repopulation of organisms.

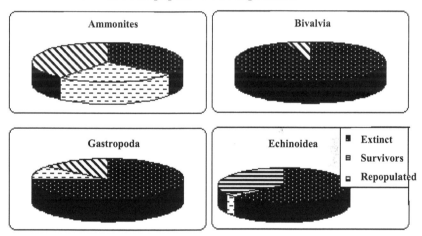

Fig. 9. Summary pie diagram depicting the diversities of Extinct, Repopulated and Survivors

9.7 Discussion and conclusions

The macrofaunal categories employed to infer paleoenvironmental conditions. Collectively the methods of establishment of the optimum conditions for the paleoecology are more precisely approached by analyzed the association grouply. We employ several lines of evidence from invertebrate fossils to propose a paleoenvironmental interpretation. The disappearance of many species and the general depletion of oyster diversity in late Cenomanian deposits are induced by the coeval sea-level rise and platform flooding that destroyed the shallow water environments. Although, some species persist during this event, and abundantly occur in some parts of the Turonian platform succession, the assemblages are less diverse and are less important as carbonate producing organisms. Cenomanian assemblages of all groups are well comparable with those of other Tethyan realm areas. Regarding of the stratigraphic distribution of these fauns revealed that the temperature, salinity, oxygenation, nutrients and water circulation is the effective variables controlled distribution of this biota.

Raha Formation deposited under lower infralittoral condition with warm and moderate oxygenation water percentage, Abu Qada Formation deposited under upper infralittoral to upper circalittoral environment with high temperature, low oxygen percentage and quiet water conditions. Uum Omieyid Formation settled in high-agitated infralittoral environment with moderate temperature and oxygen percentage. Hawashiya Formation accumulates under upper infralittoral environment with normal oxygenation in arid conditions.

Albian-Cenomanian boundary is continuous with no change in the depositional mode. Cenomanian/Turonian boundary is taken at the surface of maximum flooding, a great mass extinction for all macroinvertebrate fauna. A major hiatus at Turonian/Coniacian boundary suggest subarial exposure and represents an erosional surface at the base of Hawashiya Formation. Coniacian/Santonian boundary characterize by return to homogenous condition and flourishing of macroinvertebrate fauna again.

The overall patterns of the C/T mass extinction are characteristic for the transition in the study area. There are extinction, survival, and recovery macroinvertebrates at different durations, although their durations differ somewhat in detail, the events are also remarkably similar. The nektic organisms, which inhabited the upper water column (Ammonites), were virtually immune to extinction events. The epifaunal bivalves (*Neithea, Plicatula, Inoceramus*) were another group that was virtually unaffected or slightly affected by these events, and in both cases they became the most dominant organisms in the late portions of the extinction interval and throughout the survival interval reflecting the low-oxygen tolerant taxa which were common during greenhouse intervals. This suggests that events forced by anoxia have a predictable fabric; the biotic response is less random than has been suggested by some hypotheses, such as contingency. Finally, there is a strong link between repopulation pattern and environmental conditions. The C/T extinctions suggest that the magnitude and duration of anoxia and its associated environmental stress was at least a significant player in determining the biotic response. In addition, macroinvertebrates can be used as a predictive tool in as much as when they returned toward background levels, suggesting environmental amelioration. Comparison study with the study of Harries and Crispin 1999 revealed that significant ecological disruption of the C/T is a great marker world wide event. In both cases ammonites and epifaunal Bivalvia was unaffected by anoxia or being able to tolerate the ecological disruption. According to Harries and Crispin 1999 the biota became more diverse and gradually began to resemble pre-extinction biotas as conditions ameliorated, these evidence can't be achieved here and hence it may reflect a local changes related to the different geological setting of the studied areas.

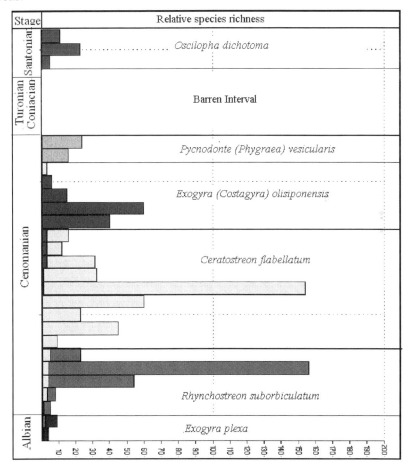

Fig. 10. Relative Species richness of Albian-Santonian index oysters

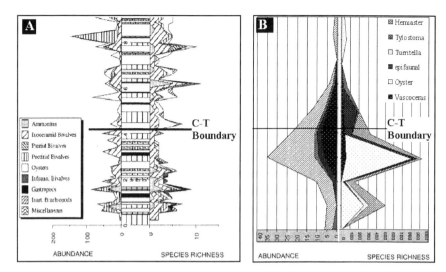

Fig. 11. Comparison of C/T mass extinction in this study with Harries and Crispin 1999

Analyzing diversity and abundance patterns of benthic macroinvertebrates organisms across the C/T boundary based on the faunal content of the North Wadi Qena, Eastern Desert, Egypt revolved the following; extinction peak is found in analysis of the raw data of stratigraphic ranges. Evidence for selective extinction is great when the analysis is limited to the boundary interval alone and when focused on taxonomic and ecological characteristics of individual genera. When taxa are separated by environmental preferences, several determinants of extinction risk become evident, suggesting that ammonites had a significantly higher extinction risk than level-bottom dwellers, much of this selectivity is not independent and also seen in the intervals of background extinctions suggesting that the C/T mass extinction represents qualitatively and quantitatively intensification of the different macro evolutionary regime.

9.8 Acknowledgements

The acknowledge goes to Prof. Dr. M. S. Ali, Dr. M. M. Azab and Dr. M. H. El-Dawy, Geology Department, Faculty of Science, El-Minia University, Egypt for helping during fieldwork. Thanks to Prof. Dr. A. S. Kassab, Geology Department, Faculty of Science, Assiut University and Dr. M. Zakhera, Geology Department, Faculty of Science, South Valley

University for helpful discussions on the stratigraphy of the C/T transition on Egypt. A special word of thanks is due to Prof. Dr. P. Kelley (USA) for critical reviewing of the manuscript.

References

Abdel-Gawad GI, El Sheikh HA, Abdelhamid MA, EL Beshtawy MK, Abed M, Fürsich FT (2004) Stratigraphic studies on some upper Cretaceous succession in Sinai, Egypt. Egypt. J Paleont 263-303

Abdelhamid MA, El Qot GM (2001) Upper Cretaceous Echinoid from Gabel El Hamra and Gabel El Minsherah, North Sinai, Egypt. MERC Ain Shams University, Earth Science 15, 1-20

Abdelhamid MA, Azab MM (2003) Aptian Cenomanian echinoids From Egypt. Review Pleobiologie, Geneve 22 (2): 851-876

Abdelhady AA (in Press) Unitary Association based biozonation; A Case study from the Upper Cretaceous strata North of Wadi Qena, Eastern Desert, Egypt. Palaios

Dhondt AV, Malchus N, Boumaza L, Jaillard E (1999) Cretaceous oysters from North Africa-origin and distribution. Bull de la Société Géologique de France 170: 67-76

El-Hedeny M, Nafee SA (2001) Upper Cenomanian ammonites from Bir Quiseib, Northern Galala, Easrern Desert, Egypt. Egypt J Paleont 1: 115-134

El-Hedeny MM, Abdel Aal AA, Maree M, Seeling J (2001) Plicatulid bivalves from the Coniacian-Santonian Matulla Formation, Wadi Sudr, western Sinai, Egypt. Cret Res 22: 295-308

Hallam A (1967) The interpretation of size-frequency distributions in molluscan death assemblages. Palaeontology 10 (1): 25-42

Haq B, Hardenbol J, Vail PR (1987) Chronology of fluctuating sea levels since the Triassic. Science 235: 1156-1167

Harries PJ (1993) Dynamics of survival following the Cenomanian –Turonian (Upper Cretaceous) mass extinction event. Cret Res 14: 563–583

Harries PJ (1999) Repopulations from Cretaceous mass extinctions: Climatic and/or evolutionary controls? Barrera E, Johnson CC (eds), The Evolution of the Cretaceous Ocean/Climate System (chemostratigraphic signals and the early Toarcian anoxic event). Geol Soc Am Spec Pap Sedimentology 44: 687–706.

Harries PJ, Kauffman EG (1990) Patterns of survival and recovery following the Cenomanian–Turonian (Late Cretaceous) mass extinction in the Western Interior Basin, United States. In Kauffman EG, Walliser OH (eds) Extinction Events in Earth History. Lecture Notes in Earth History 30, Springer, Berlin, 277-298

Harries PJ, Crispin TS (1999) The early Toarcian (Early Jurassic) and the Cenomanian–Turonian (Late Cretaceous) mass extinctions: similarities and contrasts. Palaeogeography, Palaeoclimatology, Palaeoecology 154: 39–66

Huber BT, Hodell DA, Hamilton CP (1995) Middle-Late Cretaceous climate of the southern high latitudes: stable isotopic evidence for minimal equator-to-pole thermal gradients. Geological Society of America Bulletin 107: 1164-1191

Kauffman EG (1981) Ecological reappraisal of the German Posidonienschiefer (Toarcian) and the stagnant basin model. In Gray J, Boucot AJ, Berry WBN (eds) Communities of the Past. Hutchinson and Ross, Stroudsburg, 211-381

Kauffman EG (1982) The community structure of 'shell islands' on oxygen depleted substrates in Mesozoic dark shales and laminated carbonates. In Einsele G, Seilacher A (eds) Cyclic and Event Stratification. Springer, Berlin, 502-503

Kauffman EG (1984) The fabric of Cretaceous marine extinctions. In Berggren WA, van Couvering JA (eds) Catastrophes in Earth History. Princeton University Press, Princeton, NJ, 151-246

Kauffman EG (1988) Concepts and methods of high-resolution event stratigraphy. Annul Rev Earth Planet Sci 16: 605-654

Kauffman EG (1995) Global change leading to biodiversity crisis in a greenhouse world: the Cenomanian–Turonian (Cretaceous) mass extinction. In Stanley SM, Usselmann T (eds) The Effects of Past Global Change on Life. Studies in Geophysics, National Academy Press, Washington, 47-71

Kauffman EG, Erwin DH (1995) Surviving mass extinctions. Geotimes 14: 14-17

Kauffman EG, Harries PJ (1996) The importance of crisis progenitors in recovery from mass extinction. In Hart MB (ed) Biotic Recovery from Mass Extinction Events. Geol. Soc. London, Spec Publ 102: 15-39

Kassab AS (1999) Cenomanian-Turonian boundary in the Gulf of Suez region, Egypt: towards an inter-regional correlation, based on ammonites. Geol Soc Egypt, Special Publ 2: 61- 98

Keller G, Adatte T, Stinnesbeck W, Luciani V, Karoui-Yaakoub N, Zaghbib-Turki D (2002) Paleoecology of the Cretaceous/Tertiary mass extinction in planktonic foraminifera, Paleogeography, Paleoclimatology, Paleoecology 178: 257-297

Khalil M, Meshaly S (2004) The Cretaceous / Paleogene invertebrate fauna of Gabel Mesaba Salama area, Egypt. J Paleont 4: 1-34

Klitzsch E, Hermina M (1989) The Mesozoic. In Stratigraphic lexicon and explanatory notes to the geological map of Egypt 1:500 000 (Conoco Inc, Cairo), 77-140

Kora M, Hamama HH (1987) Biostratigraphy of the Cenomanian–Turonian successions of Gebel Gunna, southeastern Sinai, Egypt. Mansoura Fac Sci Bull 14: 34-62

Krenkel E (1924) Der Syrische Bogen. Centralblatt für Mineralogie, Geologie und Palaeontologie Abhandlungen B 9, 274-281; 10: 301-131

Luger P, Groscke M (1989) Late Cretaceous ammonites from the Wadi Qena Area in the Egyptian Eastern Desert, Paleontology 32 (2): 355-407

Mancini EA (1978) Origin of the Grayson Micromorph Fauna (Upper Cretaceous) of North- Central Texas. J Paleont 52: 1294-1314

Néreaudeau D (1995) Diversité des échinides fossiles et reconstitutions paléoenvironnementales. Geobios 3: 293–324

Néreaudeau D, Thierry J, Moreau P (1997) Variation in équinoid biodiversity during the Cenomanian-early Turonian transgressive episode in Charentes (France). Bulletin de la Société Géologique de France 168 (1): 51–61

Olsson RK, Wright JD, Miller KG (2001) Paleobiogeography of *Pseudotextularia elegans* during the latest Maastrichtian global warming event. J Foram Res 31 (3): 275–282

Raup DM (1984) Evolutionary radiations and extinctions. In Holland, H.D., Trendall AF (eds) Report of the Dahlem Conference Workshop on Patterns of Change in Earth Evolution. Springer, Berlin, 5–14

Seeling J, Bengston P (1999) Cenomanian oysters from the Sergipe Basin, Brazil. Cret Research 20: 747–765

Shahar J (1994) The Syrian arc system: an overview. Palaeo3 112: 125–142

Thurow J, Brumsack H-J, Rullkötter J, Littke R, Meyers P (1992) The Cenomanian/Turonian boundary event in the Indian Ocean- A key to understand the global picture. In Duncan RA, Rea DK, Kidd RB, Rad UV, Weissel JK (eds) Synthesis of Results from Scientific Drilling in the Indian OceanAmeric. Geophysical Union, Geophysical Monograph 70: 253–273

Videt B, Platel JP (2003) Les ostréidés des faciès lignitifères du Crétacé moyen du Sud-Ouest de la France (Charentes et Sarladais). C R Pale 4: 67–176

Walker SE (2001) Paleoecology of gastropods preserved in turbiditic slope deposits from the Upper Pliocene of Ecuador. Palaeogeography, Palaeoclimatology, Palaeoecology 166: 141–163

Zakhera M (2002a) Upper Cretaceous (Cenomanian-Maastrichtian) gastropods from west of the Gulf of Suez, Egypt. Stuttgart N Jb Abh 225 (3): 297–336

Zakhera M 92002b) New record of Inoceramid bivakves from the Upper Cretaceous of Eastern Desert, Egypt. Egypt J Palontol 2: 345–357

Zakhera M (2003) Biostratigraphy and paleoecology of the Upper Cretaceous gastropod fauna from Egypt, The 3d Intern Conf Geol Africa 2: 443–462

10 K-Pg mass extinction

Ashraf M. T. Elewa

Geology Department, Faculty of Science, Minia University, Minia 61519, Egypt, aelewa@link.net

One of the mysteries of the history of the Earth is the layer of clay that was deposited around the entire globe approximately 65 million years ago. The layer marks the K-T boundary the end of the Cretaceous (K) and beginning of the Tertiary (T) periods. It is best known as the time when not only the dinosaurs, but also nearly half of all life forms became extinct (Cowen 2005).

The Wikipedia (the free encyclopedia) stated that The Cretaceous-Tertiary extinction occurred about 65.5 million years ago. It is also known as the K-T extinction event and its geological signature as the K-T boundary ("K" is the traditional abbreviation for the Cretaceous Period, to avoid confusion with the Carboniferous Period, abbreviated as "C"). Since the label "Tertiary" is no longer recognized by most geologists (for example, the International Commission on Stratigraphy) as a geologic 'Period', the K-T demise might also be called the Cretaceous-Paleogene (or K-Pg) extinction event.

Since about 65.5 Ma large vertebrates as well as most planktons and several tropical invertebrates, particularly reef-dwellers, became extinct on Earth. This abrupt extinction occurred at the end of the Cretaceous Period. At the same time, many land plants were harshly affected.

Smithsonian National Museum of Natural History (Department of Paleobiology) argued that sixty-five million years ago the curtain came down on the age of dinosaurs when a cataclysmic event led to mass extinctions of life. This interval of abrupt change in Earth's history, called the K/T Boundary, closed the Cretaceous (K) Period and opened the Tertiary (T) Period.

This deep-sea core provides convincing support to the hypothesis that an asteroid collision devastated terrestrial and marine environments worldwide. It also shows a record of flourishing marine life before the event, followed by mass extinction and then evolution of new species and slow recovery of surviving life forms after the event.

University of California (Museum of Paleontology) asked: What Killed The Dinosaurs? Then, scientists of this big university presented both valid and invalid hypothesis. The most popular valid hypothesis is that of a group of scientists at the University of California at Berkeley — Luis and Walter Alvarez, Frank Asaro, and Helen Michel — who proposed a stunning and convincing mechanism for the "K-T extinction" (meaning the extinction of dinosaurs at the boundary between the Cretaceous period (K) and the Tertiary period (T)). On the other hand, the invalid hopotheses that were presented by the museum of the University of California are as follows in types of questions:

1. Hay fever killed the dinosaurs?
2. Sniffles killed the dinosaurs?
3. Dinosaurs got so darned big that they crushed themselves?
4. Mammals outcompeted the dinosaurs?
5. Mammals ate all of the dinosaurs eggs?
6. Cosmic rays killed the dinosaurs?
7. The dinosaurs just faded away into extinction?

However, scientists clued that none of these causes led to extinction of dinosaurs. Moreover, we should search for causes led to extinction of several groups of organisms not only dinosaurs!!

We should also note that there are two main schools in this subject matter, one believes in the impact theory (e.g. Alvarez et al. 1980), and the other believes in the non-impact hypotheses (e.g. Twitchett 2006). Yet, a third school believes in multiple hypotheses for this mass extinction event (e.g. Molina et al. 1996, Dakrory 2002, Keller 2003, Keller et al. 2003, MacLeod in press, Elewa and Dakrory submitted-1, Elewa and Dakrory submitted-2).

Prothero (1998) commented on this situation by stating that many scientific papers have been rejected, grant proposals sabotaged, and careers ruined in the bare-knuckle fisticuffs over the K-Pg boundary controversy. He mentioned that the debate has become so angry and polarized that almost no evidence will change minds of the major players, because they are so committed to the positions they have argued for several years that they cannot afford to change positions and lose face as well as funding.

Nonetheless, since 1998, when Prothero mentioned his words, there are continuous trials to define the exact causes of this mass extinction, and fortunately recent studies support the third school in assigning this event to multiple causes.

References

Alvarez LW, Alvarez W, Asaro F, Michel HV (1980) Extraterrestrial cause for the Cretaceous-Tertiary extinction. Science 208: 1095-1108

Cowen R (2005) History of Life. Blackwell (4th edition)

Dakrory AM (2002) Biostratigraphy, paleoenvironment and tectonic evolution of the Late Cretaceous- Early Paleogene succession on the North African Plate (Sinai, Egypt) and a comparison with some European and Asian sections. Tuebinger Geowissen. Arbeiten (TGA), Reihe A, Band 64, 263 pp

Elewa AMT, Dakrory AM (submitted-1) Patterns and causes of mass extinction at the K/Pg boundary: Planktonic foraminifera from the North African Plate. Cretaceous Research

Elewa AMT, Dakrory AM (submitted-2) Causes of mass extinction at the K/Pg boundary: A case study from the North African Plate. Earth Science Reviews

Keller G (2003) Biotic effects of impacts and volcanism. Earth and Planetary Science Letters 215 (2003): 249-264

Keller G, Stinnesbeck W, Adatte T, Stueben D (2003) Multiple impacts across the Cretaceous-Tertiary boundary. Earth Science Reviews, 62 (2003): 327-363

MacLeod N (in press) End-Cretaceous extinctions. In Selley RC, Cocks, LRM, Plimer IR (eds) Encyclopedia of Geology. Academic Press, London. 346 pp

Molina E, Arenillas I, Arz JA (1996) The Cretaceous/Tertiary boundary mass extinction in planktic foraminifera at Agost, Spain. Rev Micropaléont 39 (3): 225-243

Prothero DR (1998) Bringing fossils to life: An introduction to paleobiology. WCB/McGrow-Hill, USA, 560 pp

Twitchett RJ (2006) The palaeoclimatology, palaeoecology and palaeoenvironmental analysis of mass extinction events. Palaeogeogr, Palaeoclimat, Palaeoecol 232 (2006): 190-213

11 Causes of mass extinction at the K/Pg boundary: A case study from the North African Plate

Ashraf M. T. Elewa and Ahmed M. Dakrory

Geology Department, Faculty of Science, Minia University, Egypt, aelewa@link.net

11.1 Introduction

"Humans are responsible for the worst spate of extinctions since the dinosaurs and must make unprecedented extra efforts to reach a goal of slowing losses by 2010"

This sentence was mentioned in the UN report dated 3/21/2006. However, are the causes same as those resulted in the five major mass extinctions of the fossil record? The answer is, of course, no. Then, what are the causes led to mass extinctions in the fossil record? We herein try to answer this question through the study of the K/Pg boundary of the North African Plate.

Twitchett (2006) stated that in the past 25 years the study of the five major mass extinctions of the fossil record (Late Ordovician event, Late Devonian event, Late Permian event, Late Triassic event and the Cretaceous/Paleogene event or K/Pg event) has increased dramatically, with most focus being on the K/Pg event. He added that many aspects of these five events are still debated and there is no common cause or single set of climatic or environmental changes common to these events, although all are associated with evidence for climatic change.

Although the causes of mass extinction at the K/Pg boundary seem to be, somewhat, different from those resulted in the other four extinctions, but still there are similar causes like global climate change related to volcanic activities, sea level changes, and fluctuation of environmental factors (e.g. productivity, oxygenation and temperature). There are two major hypotheses for the K/Pg mass extinction event. The first one suggests that the event is a more progressive and multi-causal series of events resulting from a combination of environmental and climatic factors during the latest Maastrichtian including rapid warming followed by abrupt cooling during the last 400 kyr of the Maastrichtian (e.g. Kauffman

1984; Li and Keller 1998a, b; Keller 2002). The second one points to an extremely brief worldwide catastrophe (e.g. Alvarez et al. 1980; Thierstein 1982; Smit et al. 1992; Liu and Olsson 1992; Molina et al. 1998).

Unfortunately, most studies on mass extinction at the K/Pg boundary in Egypt stress on lithostratigraphical and biostratigraphical characteristics and neglect the geochemical analyses of carbon and oxygen isotopes. This could be due to the lack of labs prepared for these techniques in most universities and research institutes in Egypt. However, collaborations with foreign institutes in Europe and USA, and elsewhere could facilitate these procedures in the last decade. Hence, the collaboration between the Geology Department at Minia University of Egypt with the Institute and Museum of Paleontology at Tuebingen University of Germany resulted in the fruitful researches in the subject through the Ph. D. of Dr. Dakrory (second author of the present paper) as well as this present study.

11.2 Material and methods

For the purpose of the present study, we selected thirty samples across the K/Pg boundary of three surface sections from Sinai in Egypt. One section is located in Northern Sinai (section Sahaba) and two sections are located in west Central Sinai (sections Hiala and El Seig) (Fig. 1). These three sections have been selected for geochemical analyses because of their suitability for such analyses through well-preserved faunal assemblages as well as containing the most complete record of the K/Pg boundary interval (see Dakrory, 2002). The geochemical analyses (carbon and oxygen isotopes, concentrations of Iridium and other siderophile elements, and concentrations of the rare earth elements) used in this study were made by Dakrory during his stay at Tuebingen University for completion of his Ph. D. Thesis. For more details see Dakrory (2002).

The strategy of using the techniques for the purpose of the present study was made in the following manner. Firstly, cluster analysis based on the Euclidean distance measure of similarity (the weighted paired group method) was applied to the raw data, to find out any clustering within samples resulting from carbon and oxygen isotope measures of the whole rock samples as well as selected foraminiferal taxa. Secondly, principal coordinate analysis based on the Gower's index of similarity was applied to the raw data to define the importance of carbon and oxygen isotopes and the selected taxa in the distribution of the foraminiferal assemblages of the study area. Thirdly, relations between the defined important taxa and the carbon and oxygen isotopes were conducted through biplots to clarify the manner of the mass extinction occurred at the K/Pg boundary of the studied sections. Finally, data obtained by Dakrory (2002) on Iridium anomaly and rare earth elements have been combined with results attained from the mentioned techniques to define the causes of the K/Pg mass extinction in the study area.

The software used for quantitative analyses and biplots is PAST, version 1.14, by Oyvind Hammer and D. A. T. Harper (September, 2003; website: http://folk.uio.no/ohammer/past). For details on PAST software see Hammer et al. (2001).

11.3 Quantitative results

Because qualitative results are based on quantitative results in our analyses, therefore it is appropriate to start with quantitative results before mentioning qualitative observations.

11.3.1 Cluster analysis

Cluster analysis based on the Euclidean distance measure of similarity (the weighted paired group method) applied to the thirty selected samples across the boundary resulted in two main groups (Fig. 2). Group "A" with samples below the K/Pg boundary, which displayed high carbon $\delta^{13}C$ content for the whole rock samples. Group "B" with samples at and above the boundary with low carbon $\delta^{13}C$ content for the whole rock samples.

The analysis revealed that the carbon $\delta^{13}C$ content (productivity) was the most effective factor on the distribution of faunal assemblages of the study area. Whereas, oxygen $\delta^{18}O$ content (e.g. oxygenation and temperature) did not display this strong effect like carbon, hence the clustering (Fig. 2) is based on the carbon content of these samples rather than their oxygen content. Because of lacking suitable information on the oxygen content from cluster analysis, we applied the principal coordinate analysis in the next step.

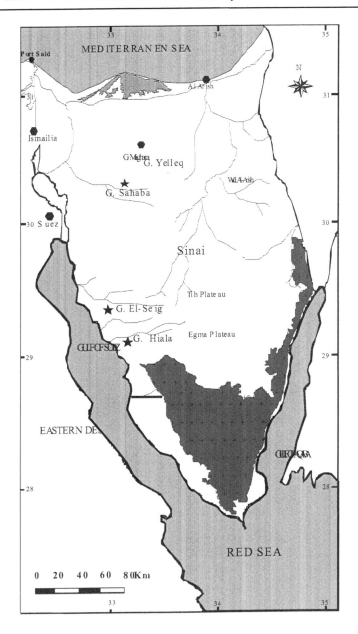

Fig. 1. Location map of the studied sections from Sinai, Egypt (after Dakrory 2002). Stars refer to the studied sections

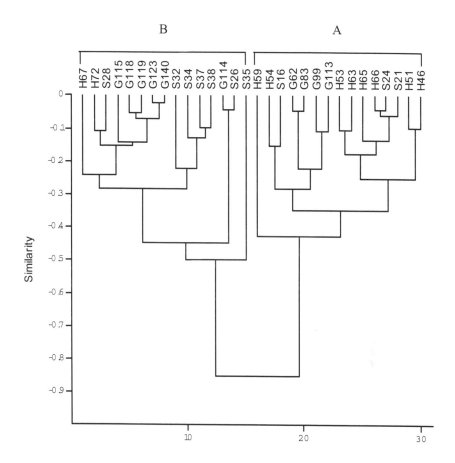

Fig. 2. Dendrogram resulting from cluster analysis of thirty samples across the K/Pg boundary of the three studied sections

11.3.2 Principal coordinate analysis

The data matrix used in cluster analysis was treated by principal coordinate analysis. The results are summarized in Table 1. This table shows that the first three latent roots account for more than 50% of the trace, which seems suitable for interpreting the important causes led to mass extinction in the study area.

Table 1. Summary of principal coordinate analysis

Eigenvalues	Value	% Variance
1	4.386	29.337
2	1.906	12.751
3	1.254	8.392

Figure 3, plotting the first vs. second principal coordinates, separates the thirty studied samples according to their carbon $\delta^{13}C$ content of *Gumbelitria cretacea* (PCO1) and their oxygen $\delta^{18}O$ content of *Cibicides* spp. (PCO2). This figure divides the studied samples into two main groups (A, B). Group "A" stands for samples below the K/Pg boundary. It shows high carbon content for *Gumbelitria cretacea* and medium carbon content for *Cibicides* spp. Group "B" represents samples at and above the K/Pg boundary. This group could be subdivided into two subgroups according to the oxygen content. Subgroup "B1" signifies samples with lower oxygen content for *Cibicides* spp., and subgroup "B2" corresponds to samples with higher oxygen content for *Cibicides* spp. (note that the positive and negative signs have no meaning in the plot). It is also obvious, from the graph, that group "A" shows lower oxygen content than group "B2".

Fig. 4, plotting the second vs. third principal coordinates, reveals that the third coordinate (PCO3) separates the studied samples according to the oxygen $\delta^{18}O$ content of *Gumbelitria cretacea*. It is clear from the plot that specimens of *Gumbelitria cretacea* exhibit relatively lower oxygen values for samples of section Sahaba (S) in the north, and higher oxygen values for samples of sections Hiala (H) and El Seig (G) in the south.

11.4 Qualitative results

Qualitative observations of the studied samples indicate the following results:

The carbon content, using isotope analysis of representatives of the whole rock samples across the K/Pg boundary of the three studied sections, decreases northward with clear excursion at samples H67, G114 and S26; these samples are located at the K/Pg boundary (Fig. 5).

All samples at and above the K/Pg boundary, except S35, show clear drop in carbon content using isotope analysis of representatives of *Gumbelitria cretacea* specimens across the K/Pg boundary of the three studied sections (Fig. 6).

Samples at the K/Pg boundary show almost the lowest values of oxygen content, using isotope analysis of representatives of *Cibicides* spp. specimens across the K/Pg boundary of the three studied sections. However at section G (G. El Seig), samples just above the K/Pg boundary (e.g. G115, G118) show lower oxygen values than the sample G114 at the K/Pg boundary (Fig. 7).

Scatter plot for the oxygen excursion $\delta^{18}O$ using isotope analysis of representatives of *Cibicides* spp. specimens across the K/Pg boundary of the three studied sections indicates that section G displays the lowest oxygen content of the three studied sections. While, section S (Sahaba) reveals higher oxygen values, and section H (Hiala) discloses medium oxygen values (Fig. 8).

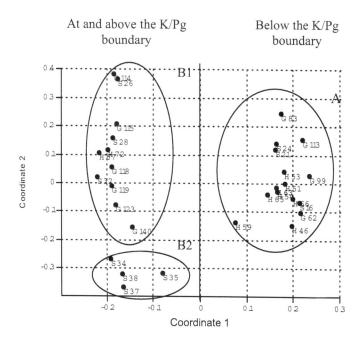

Fig. 3. Plot of the 1st versus 2nd principal coordinate axes for the studied samples across the K/Pg boundary of the three studied sections

Our observations signify that *Gumbelitria cretacea* ranged from abundant (> 18 %) below the K/Pg boundary to frequent (10-18 %) at and above the boundary in section Hiala (in the south). However, its abundancy remained below and above the boundary in section El Seig (in the north of section Hiala). Northwards, this species showed convergent strategy in section Sahaba, where it ranged from frequent below the boundary to abundant at and above the boundary. Surprisingly, the carbon content for this species showed higher values below the boundary and lower values at and above the boundary. This result indicates that *Gumbelitria cretacea* became more abundant in section Sahaba with decreasing the values of carbon $\delta^{13}C$ content.

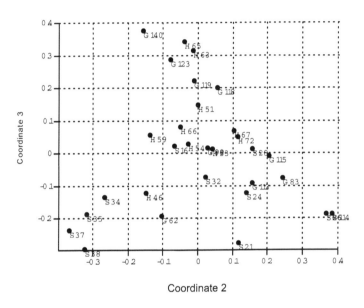

Fig. 4. Plot of the 2nd versus 3rd principal coordinate axes for the studied samples across the K/Pg boundary of the three studied sections

The last note suggests that there are two morphotypes within this species, one disappeared below the boundary, and the second could survive above the boundary. From our observations we found that the Lilliput effect is prominent in the foraminiferal assemblages of the studied sections. Therefore, *Gumbelitria cretacea* is represented in the material by two sizes, the bigger below the boundary and the smaller at and above the boundary. Consequently, the *Gumbelitria cretacea* Zone (P0) of previous studies is named as *Gumbelitria cretacea* morphotype "B" Zone in the present study.

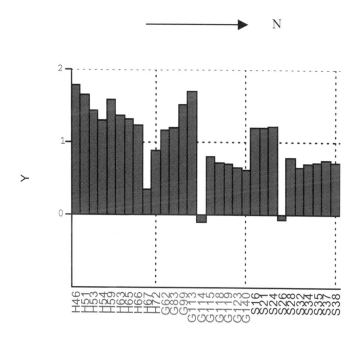

Fig. 5. Bar chart for the carbon excursion ($\delta^{13}C$) using isotope analysis of representatives of the whole rocks across the K/Pg boundary of the three studied sections (note that the carbon content decreases northward with clear excursion at samples H67, G114 and S26; these samples are located at the K/Pg boundary)

The smaller representative of *Gumbelitria cretacea* (morphotype B) denotes to a survival strategy attempt at the K/Pg boundary. Bassiouni and Elewa (1999) observed a similar strategy for the ostracod species *Paracosta mokattamensis praemokattamensis*, which was considered by them as a good stratigraphic indicator of the lower/middle Eocene boundary of Egypt.

Gradual faunal change in the studied foraminiferal assemblages indicates the occurrence of gradual mass extinction in the study area.

The measured iridium anomaly values and rare earth elements indicate both gradual and catastrophic extinctions (Dakrory 2002).

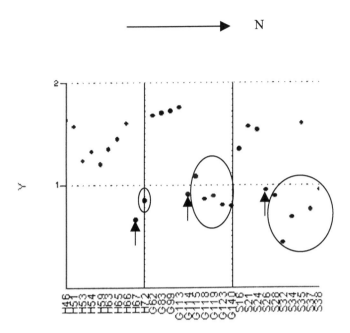

Fig. 6. Scatter plot for the carbon excursion ($\delta^{13}C$) using isotope analysis of representatives of *Gumbelitria cretacea* specimens across the K/Pg boundary of the three studied sections (vertical arrows refer to samples located at the K/Pg boundary; ellipses contain samples above the K/Pg boundary). Note that this analysis shows more details than that of analyzing the whole rocks. All samples at and above the K/Pg boundary, except S35, show clear drop in carbon content for the three studied sections

Studying the lateral changes in facies in the area of study indicates that this change from marl in the south (Gebel Hiala) to argillaceous limestone (Gebel El Seig), to chalk in the north (Gebel Sahaba) was accompanied by a distinct increase in the degree of impact in the same direction at the K/Pg boundary of the three studied sections.

11.5 Discussion and conclusions

Since the paper of Alvarez et al. (1980), which was published in Science, the subject of delineating the causes of the K/Pg mass extinction received several papers with different arguments.

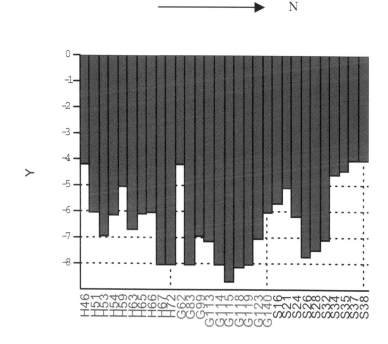

Fig. 7. Bar chart for the oxygen excursion ($\delta^{18}O$) using isotope analysis of representatives of *Cibicides* spp. specimens across the K/Pg boundary of the three studied sections. Samples at the K/Pg boundary show nearly the lowest values of oxygen content for the three studied sections. However at section G (G. El Seig), samples just above the K/Pg boundary (e.g. G115, G118) show lower oxygen values than the sample G114 at the K/Pg boundary

After more than twenty years of that pioneer paper, Keller (2003) and Keller et al. (2003) are still convinced that separating the biotic effects of impacts from those of volcanism is difficult, if not impossible. They mentioned that this task is further complicated by the recent discoveries of multiple impacts across the K/Pg boundary.

However, the combination of qualitative and quantitative observations might give useful information for understanding and interpreting the causes of the K/Pg mass extinction of the North African Plate, as a case study that could be generalized to other regions of low latitude.

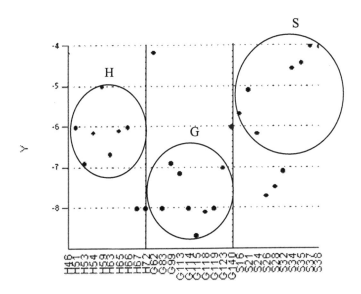

Fig. 8. Scatter plot for the oxygen excursion ($\delta^{18}O$) using isotope analysis of representatives of *Cibicides* spp. specimens across the K/Pg boundary of the three studied sections. The plot shows that section G (El Seig) displays the lowest oxygen content of the three studied sections, whereas section S (Sahaba) reveals higher oxygen values, and samples of section H (Hiala) disclose medium oxygen values

Our analyses revealed that *Gumbelitria cretacea* and *Cibicidoides* spp. are the best to clarify the carbon and oxygen excursions across the K/Pg boundary. Keller (2003) argued that preliminary $\delta^{13}C$ species ranking suggests that *Gumbelitria* is one of the isotopically lightest species living in the upper few meters of surface waters, and thrived in eutrophic waters. She added that further geochemical studies, including stable isotope and trace element analyses, are required on *Gumbelitria* tests to investigate toxicity levels that proved lethal for specialized species, as well as nutrients, salinity, temperature and oxygen variations. Dakrory (2002) concluded that *Gumbelitria cretacea* is a near surface cosmopolitan species. This species together with other species of *Gumbelitria* (e.g. *Gb.*

trifolia and *Gb. Danica*) could survive across the K/Pg boundary. Whereas, the severe conditions at the K/Pg boundary affected the tropical to subtropical deep-sea foraminiferal species. These results indicate that sea level fall should have a strong effect on the deep-sea foraminiferal taxa. Productivity (PCO1), as a result, might have disturbed according to this eustatic lowering of the sea level.

On the other hand, Elewa et al. (2004) and Kaiho (1991, 1992) stated that epifaunal foraminiferal species like *Cibicidoides* spp. and others (aerobic forms of Bernhard, 1986) prefer to inhabit environments with good oxygen conditions.

Cluster analysis could successfully separate the studied samples according to the carbon content of the whole rock samples, nevertheless an additional technique was necessary to spot the light on the oxygen content of our data and to shear any diagenetic factor that might disturb the results. Therefore, the principal coordinate analysis was applied to the same data matrix. This technique could add new information and define the best representatives of carbon and oxygen excursions in the studied samples in the following manner.

Thirty samples across the K/Pg boundary of three surface sections from North and west Central Sinai, representing the North African Plate, have been subjected to quantitative analyses of their carbon $\delta^{13}C$ and oxygen $\delta^{18}O$ contents. Observations resulted from the combination of both quantitative and qualitative analyses led to the following conclusions:

Gumbelitria cretacea is the best representative for the carbon $\delta^{13}C$ excursion (see PCO1 of Fig. 3) in the study area, Whereas, *Cipicides* spp. are the best representatives for measuring the oxygen $\delta^{18}O$ excursion (see PCO2 of Fig. 3). Moreover, *Gumbelitria cretacea* is considered, in the present study, to be a good representative of the oxygen $\delta^{18}O$ excursion, subsequent to the *Cipicides* spp. (see PCO3 of Fig. 4).

Gebel Sahaba at Northern Sinai was the most affected section within the three studied sections by the global change occurred at the K/Pg boundary;

A distinct facies bias was recognized in patterns of extinction of foraminiferal assemblages of the study area. Lateral changes in facies from marl in the south (Gebel Hiala) to argillaceous limestone (Gebel El Seig), to chalk in the north (Gebel Sahaba) are associated with distinct increase in the degree of impact in the same direction at the K/Pg boundary of the three studied sections.

Gumbelitria cretacea should have two morphotypes with different sizes. Morphotype "A" is below the K/Pg boundary, and morphotype "B", with smaller (dwarfed) specimens, at and above the boundary. Consequently, the *Gumbelitria cretacea* Zone (P0) of previous studies is defined here as *Gumbelitria cretacea* morphotype "B" Zone.

Mass extinction of the studied fauna shows clear selectivity exposure to test size. The Lilliput effect is clear within the studied foraminiferal assemblages across the boundary.

It seems that productivity due to sea level fall was the most effective factor on the distribution of foraminiferal and nannoplankton assemblages across the K/Pg boundary of the study area. Oxygenation and temperature were subsequent in their importance. These mentioned factors show abrupt negative shifts in their values at the K/Pg boundary of the studied sections, and are consistent with the impact of a large meteorite causing severe changes in the atmosphere environment followed by worldwide mass extinction.

Gradual faunal change in foraminifera together with iridium anomaly values as well as rare earth elements indicate that climate change should not be neglected as one of the important causes of this mass extinction in the study area.

In summary, multiple causes (extraterrestrial bolide impact and environmental changes) led to the mass extinction in the study area at the K/Pg boundary:

1. Extraterrestrial bolide impact;
2. Low productivity exposure to sea level fall (PCO1);
3. Low oxygenation (PCO2);
4. Cooling of water (PCO3);
5. Increasing acidity of water (results from Iridium anomaly and rare earth elements).

These results coincide with the conclusions of Molina et al. (1996).

We hope that we could answer, in the present work, questions regarding the causes of the K/Pg mass extinction in respond to Molina et al. (1996), who elucidated the urgent need of further multidisciplinary studies for understanding and interpreting these causes more accurately. We also hope that we ended the argument between impactors and gradualists (see Prothero 1998), where the impactors ignore the evidence of land reptiles and amphibians, no matter how compelling it is, and gradualists have also been stubborn and hard to convince when the evidence such as stishovite and Chicxulub crater finally proved the impact had occurred.

11.6 Acknowledgements

The authors are much indebted to Prof. Dr. Hanspeter Luterbacher of the Museum of Paleontology, Barcelona University, Spain for his continuous help as well as reading the first draft of this paper. We would also thank Prof. Dr. Koebrel of the Institute of Geochemistry, University of Vienna, Austria, for doing the rare earth elements analysis in his lab. A special word of thanks is due to Prof. Dr. Satir of the Institute and Museum of Paleontology, Tuebingen University, Germany, for isotope analysis of bulk and whole rock samples and Mrs. Dora Wirgate of Michigan University, USA, for isotope analysis of selected species. Dr. Oyvind Hammer of the Natural History Museum, University of Oslo, Norway, is deeply acknowledged for explaining some of the techniques used in his PAST software.

References

Alvarez LW, Alvarez W, Asaro F, Michel HV (1980) Extraterrestrial cause for the Cretaceous-Tertiary extinction. Science 208: 1095-1108

Bassiouni MA, Elewa AMT (1999) Studies on some important ostracod groups from the Paleogene of Egypt. Geosound, Turkey, 35: 13-27

Bernhard JM (1986) Characteristic assemblages and morphologies of benthic foraminifera from anoxic, organic-rich deposits: Jurassic through Holocene. J Foraminiferal Res 16: 207-215

Dakrory AM (2002) Biostratigraphy, paleoenvironment and tectonic evolution of the Late Cretaceous- Early Paleogene succession on the North African Plate (Sinai, Egypt) and a comparison with some European and Asian sections. Tuebinger Geowissen. Arbeiten (TGA), Reihe A, Band 64, 263 pp

Elewa AMT, Dakrory AM, Omar AA, Osman, OA (2004) Foraminiferal biostratigraphy and paleoecology of the Upper Cretaceous-Lower Paleogene succession at Safaga area, Eastern Desert, Egypt. Bull Fac Sci, Assiut Univ, 33 (2-F): 15-68

Hammer O, Harper DAT, Rayan PD (2001) PAST: Palaeontological statistics software package for education and data analysis. Palaeontologia Electronica 4 (1): 9 pp

Kaiho K (1991) Global changes of Paleogene aerobic/anaerobic benthic foraminifera and deep-sea circulation. Palaeogeogr, Palaeoclimat, Palaeoecol 83 (1991): 65-85

Kaiho K (1992) A low extinction rate of intermediata-water benthic foraminifera at the Cretaceous/Tertiary boundary. Marine Micropaleont 18 (1992): 229-259

Kauffman EG (1984) The fabric of Cretaceous marine extinction. In Berggren WA, Van Couvering JA (Eds). Catastrophes and Earth History. Princeton Univ Press, 151-246

Keller G (2002) *Gumbelitria*-dominated late Maastrichtian planktic foraminiferal assemblages mimic early Danian in central Egypt. Marine Micropaleont 47 (2002): 71-99

Keller G (2003) Biotic effects of impacts and volcanism. Earth and Planetary Science Letters 215 (2003): 249-264

Keller G, Stinnesbeck W, Adatte T, Stueben D (2003) Multiple impacts across the Cretaceous-Tertiary boundary. Earth Sci Rev 62 (2003): 327-363

Li, L., Keller, G., 1998a. Maastrichtian climate, productivity and faunal turnovers in planktic foraminifera in South Atlantic DSDP Site 525A and 21. Marine micropaleont., 33: 55-86.

Li L, Keller G (1998) Diversification and extinction in Campanian-Maastrichtian planktic foraminifera of Northwestern Tunisia. Ecologae Geol. Helvetiae 91: 75-102

Liu C, Olsson RK (1992) Evolutionary radiation of microperforate planktonic foraminifera following the K/T mass extinction events. J Foram Res 22 (4): 328-346

Molina E, Arenillas I, Arz JA (1996) The Cretaceous/Tertiary boundary mass extinction in planktic foraminifera at Agost, Spain. Rev Micropaléont 39 (3): 225-243

Molina E, Arenillas I, Arz JA (1998) Mass extinction in planktic foraminifera at the Cretaceous/Tertiary boundary in subtropical and temprate latitudes. Bull Soc Géol France 169: 351-363

Prothero DR (1998) Bringing fossils to life: An introduction to paleobiology. WCB/McGrow-Hill, USA, 560 pp

Smit J, Montanari A, Swiburne NHM, Alvarez W, Hildebrand AR, Margolis SV, Claeys Ph, Lowrie W, Asaro F (1992) Tektite-bearing, deep-water clastic unit at the Cretaceous-Tertiary boundary in Northeastern Mexico. Geology 20: 99-103

Thierstein HR (1982) Terminal Cretaceous plankton extinctions: A critical assessment. Geol Soc America, Special Paper, 190: 385-399

Twitchett RJ (2006) The palaeoclimatology, palaeoecology and palaeoenvironmental analysis of mass extinction events. Palaeogeogr, Palaeoclimat, Palaeoecol 232 (2006): 190-213

12 Patterns and causes of mass extinction at the K/Pg boundary: Planktonic foraminifera from the North African Plate

Ashraf M. T. Elewa and Ahmed M. Dakrory

Geology Department, Faculty of Science, Minia University, Minia, Egypt, aelewa@link.net

12.1 Introduction

By far the most interest and attention has been focused on the great extinction that ended the Mesozoic. The obvious reason is that this event wiped out the dinosaurs. A second reason is the evidence of an extraterrestrial bolide impact (See Prothero 1998).

For a long period of time causes and patterns of mass extinction at the K/Pg boundary received conflicting results. Some assigned this mass extinction to catastrophic effects, and some others referred to gradual environmental and climatic changes. However, since Molina et al. (1996) concluded that there were multiple causes to this mass extinction, the search in the subject became more reliable and accurate (e.g. Keller 2003; Macleod in press).

Molina et al. (1996) stated that the catastrophic pattern of extinction at the K/Pg boundary is very compatible with the effect of a large meteorite impact, whereas the gradual and extended pattern of extinction across the Maastrichtian-Danian transition is compatible with temperature and sea level changes that may be related to massive volcanism.

Keller (2003) argued that the unequivocal connection between intense volcanism and high stress assemblages during the late Maastrichtian to early Danian, and the evidence of multiple impacts, necessitates revision of current impact and mass extinction theories.

Macleod (in press) cued that three prominent single-cause mechanisms have been popular in accounting for the K/Pg mass extinction, including sea-level change, a large igneous province volcanic eruption on what is now the Indian sub-continent, and the impact of a ~10 km bolide on Mexico's Yucatan Peninsula. He added that the ecological complexity of the end-Cretaceous extinctions, the time over which they took place, and the record of historical association between these mechanisms and extinctions over the last 250 m.y. suggests that no single mechanism can reasonably account for the patterns seen in the fossil record unless that record is assumed to be so strongly biased that the basis for recognizing the Maastrichtian as a time of widespread extinction is itself called into question.

On the other hand, Twitchett (2006) contradicted the supposed extinction-causing environmental changes resulting from extraterretrial impact.

In light of these published works, we attempt to explicate the patterns and causes of planktonic foraminiferal mass extinction at the K/Pg boundary of the successions located in Northern and west Central Sinai, with the aid of two multivariate data analysis techniques (hierarchical cluster analysis and neighbor joining clustering).

12.2 Stratigraphy

There are two formations that have been recorded in the strata of the four studied sections; these are, from older to younger:

12.2.1 Sudr Chalk

Ghorab (1961) introduced the term Sudr Chalk for the snow-white chalk and argillaceous limestone sequence exposed in Wadi Sudr, west Central Sinai. He further subdivided it into the lower Markha Member and the upper Abu Zenima Member and assigned a Campanian age to the Markha Member and a Maastrichtian age to the Abu Zenima Member.

In our study area, the Sudr Chalk is distributed in the lowlands between the prominent structural highs. It conformably overlies the Matulla Formation and underlies the Dakhla Formation. The upper part of this formation (Abu Zenima Member) is present in all studied sections, while the lower part (Markha Member) is missing in section Sahaba (S).

The Abu Zenima Member, which is represented in all studied sections, consists of well-bedded, yellowish-white chalky limestones. This limestone becomes more argillaceous at the top of the member. Abu Zenima Member attains 56 m in section Yelleq, 5 m in section Sahaba, 16 m in section Hiala and 45 m in section El Seig.

It is worthy to mention that section Sahaba is lithologically different from the other studied sections in Sinai. There is no clear lithological difference between the Sudr and Dakhla formations in the field and the majority of this section consists of marly and argillaceous limestones.

Moreover, there is no typical Dakhla Formation in the Sahaba and El-Seig sections.

12.2.2 Dakhla Formation

The term Dakhla Formation was introduced by Said (1961). Its type locality is located at the north of Mut, the main village of the Dakhla Oasis in the Western Desert of Egypt. Its type section was described by Said (1961) as composed of thin bone beds, phosphatic bands, and allochthonous limestones at the base, grading upward into massive shales. The total thickness of this section is 230 m.

In the study area, the Dakhla Formation is of late Maastrichtian to early Paleocene ages, and consists of greenish gray, partially glauconitic calcareous shales, and marly limestones. It conformably overlies the Sudr Chalk Formation and underlies the Tarawan Formation.

12.3 Material and methods

Forty samples across the K/Pg boundary of four sections from Sinai (Fig. 1), representing the North African Plate, have been subjected to two clustering techniques to elucidate the patterns and causes of planktonic foraminiferal mass extinction in this important area of North Africa. These two methods are the hierarchical unweighted pair-group average algorithm and the neighbor joining algorithm (both based on the Euclidean distance measure in the present study). The neighbor joining was originally made for phylogenetic analysis, but it really is just another type of agglomerative clustering. Therefore, it is possible to use the neighbor joining clustering to ensure the results obtained from hierarchical cluster analysis.

The parameters used in our data matrix are: cosmopolitan spp., tropical-subtropical spp., Paleogene spp., first appeared spp., last appeared spp., survivor spp., surface water spp., intermediate water spp., deep-water spp., heterohelicids (biserial, triserial, and multiserial forms), and trochospiral forms (globigerinids, rugoglobigerinids, and globotruncanids). Therefore, the data matrix is formed of 40 samples (in rows) and 15 parameters (in columns). The analyses were made on the rows (samples).

The dendrogram resulted from cluster analysis was created using STATISTICA software for windows, release 4.5, StatSoft Inc. (1993).

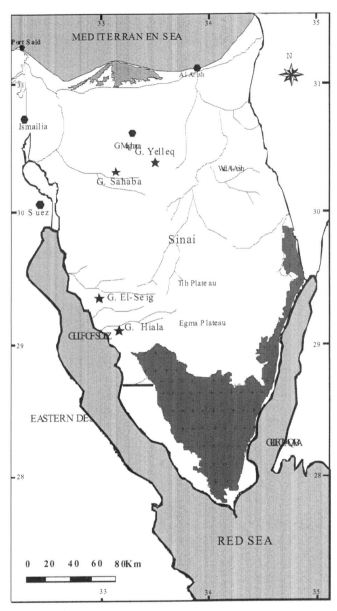

Fig. 1. Location map of the studied sections from Sinai, Egypt (after Dakrory 2002). Stars refer to the studied sections

By comparing the results with those of PAST program (Hammer et al. 2001), we found almost the same results, however we used the dendrogram that was created by STATISTICA because it displays more details. The neighbor joining clustering was created using PAST version 1.55 (2006).

12.4 Results and discussion

Most palaeontologists favor a multiple-cause explanation for the end-Cretaceous extinctions, though nonpaleontological geologists continue to favor the bolide impact model (Macleod in press). From the works of Macleod and other geologists, it is clear that the patterns and causes of the K/Pg mass extinction are still debated.

In the present paper we could successfully investigate the patterns of the K/Pg planktonic foraminiferal mass extinction in North and west Central Sinai, which belongs to the North African Plate, as a step to understanding all organismal extinctions that occurred in the uppermost interval of the Maastrichtian stage of the Cretaceous Period/System.

It is noteworthy to mention that comparing between UPGMA and neighbor joining for other types than phylogenetic data gave attractive results with additional information gained from neighbor joining clustering.

As we see from Figs. 2, 3 there are six (A-F) clusters or groups, most of the clusters are same using the two methods, which shows that the groups are robust. Nevertheless, group C is subdivided into C1, C2 in the neighbor joining clustering. It seems that we have quite clear, robust clusters (most of the within-cluster distances are small and the across-cluster distances large).

From the dendrogram (Fig. 2) we could discriminate between six groups that are interpreted as follows:
1. Group "A", with samples located below the boundary of section Hiala at the south. These samples contain high abundance of the planktonic foraminiferal species belonging to the ecological specialists (e.g. globotruncanids and rugoglobigerinids) of Keller (2002). She described these ecological specialists as tropical-subtropical species with narrow tolerance limits and restricted geographical range. All went extinct at or near the K/Pg boundary. Consequently, we conclude that group "A" was dominated by species that lived in normal marine conditions.
2. Group "B", with samples located above the boundary of section Hiala at the south. This group contains abundant heterohelicids (e.g. ecological generalists of Keller (2002)). The ecological generalists according to Keller (2002) are tolerant taxa with global

biogeographic range, indicating unstable conditions above the boundary of section Hiala.

3. Group "C", includes samples located above the K/Pg boundary of the other studied sections (sections El Seig, Sahaba and Yelleq). Globigerinids (ecological specialists) and heterohelicids (ecological generalists) are almost equal in abundance in this group, indicating starting the interval of foraminiferal recovery above the boundary.

4. Group "D", contains samples located above the boundary of sections El Seig, Sahaba and Yelleq. However, these samples are located above those samples of group "C". We noticed that globigerinids became dominant in comparison to heterohelicids in samples of group "D", pointing to the recovery interval that started in samples of group "C".

5. Group "E", with samples below the boundary at sections El Seig, Sahaba and Yelleq. This group has higher percent of the severely suffered deep-water forms than group "A", indicating trouble marine conditions.

6. Group "F", with the samples located immediately below the boundary of sections El Seig, Sahaba and Yelleq. The assemblage of these samples contain cosmopolitan together with tropical-subtropical species, and refers to the start of the environmental stress that reached its maximum at the boundary.

Note that:
1. Group "A" is different from other groups because it contains the highest total percent of the ecological specialists (globotruncanids and rugoglobigerinids). In spite of this, globotruncanids of almost intermediate water habitats are more abundant in this group than rugoglobigerinids of surface water habitats.

2. Groups "C, D" are closer to each other in the dendrogram than to group "B" as a result of the similar environmental conditions of samples related to groups "C, D" at and above the boundary of sections El Seig (G), Sahaba (S) and Yelleq (Y) than that in group "B" of section Hiala (H). The difference in the environmental conditions resulted from the greater abundance of surface species in section (H), at and above the boundary, than in sections (G, S, Y). Increasing numbers of surface species points to lesser stress than for deep-water species (see Dakrory 2002).

3. Groups "E, F" are close to each other in the dendrogram because they contain samples located below the boundary with

many similar species. However, group "F" differs from group "E" in having more abundance of extinct species just below the boundary.

4. Group "D" encloses many Paleogene species, referring to the recovery interval above the K/Pg boundary. This group is mostly formed of samples belonging to section Yelleq, showing that the recovery began in section Yelleq before the other sections.

From the tree (Fig. 3), we noticed that group "C" of the dendrogram (Fig. 2) is subdivided into two subgroups "C1, C2". "C1" is totally included within samples above the boundary in section Sahaba. It contains greater number of Paleogene species than C2, which is composed of samples at the boundary in sections El Seig, Sahaba and Yelleq, adding to samples above the boundary in section El Seig. That refers to the rapid foraminiferal recovery in section Sahaba when compared to section El Seig. In conclusion, foraminiferal recovery began earlier in the sections of Northern Sinai than those of west Central Sinai.

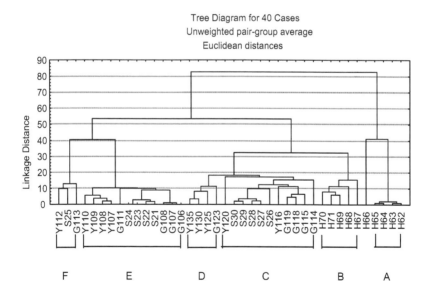

Fig. 2. Dendrogram resulting from cluster analysis of forty samples across the K/Pg boundary of the four studied sections

The established results led to the following conclusions:

1. The extraterrestrial bolide impact played an important role in the foraminiferal extinction in section Hiala (in the south) than the other three sections.
2. Sections El Seig, Sahaba and Yelleq faced gradual foraminiferal extinction earlier below the boundary due to the change in environmental and climatic conditions, and then the bolide impact together with these conditions affected the faunal assemblages at and above the boundary.
3. The relative abundance of surface species in the samples of the study area led to conclude that stress was stronger at the boundary in sections Sahaba and Yelleq in the north than sections Hiala and El Seig in the south.
4. Foraminiferal recovery started earlier in the north than in the south of the study area.

In summary, quantitative faunal analysis, using two clustering techniques, of forty samples across the K/Pg boundary from four sections in North and Central Sinai (the North African Plate) enabled us to investigate the patterns and causes of planktonic foraminiferal mass extinction at that boundary. The results revealed that both abrupt and gradual extinctions might have occurred, however patterns of these two extinctions are relatively different between the four studied sections. For example, the stress started immediately at the boundary in section Hiala (in the south), pointing to the direct effect of the extraterrestrial bolide impact on the origination and extinction of the foraminiferal fauna of this succession. Nonetheless, we cannot ignore the gradual extinction pattern that occurred simultaneously with the abrupt extinction in this section at and above the boundary. In the other sections, the stress started with gradual extinction below the boundary, indicating the strong effect of regional environmental and climatic changes in this interval. At the boundary of these sections, the bolide impact together with environmental and climatic changes led to the foraminiferal mass extinction. Moreover, our investigations show that sections Yelleq and Sahaba (in the north) suffered severely from the stress at the boundary than the other two sections (e.g. Hiala and El Seig in the south). On the other hand, it can be concluded that the faunal recovery has started earlier in the north than in the south of the study area.

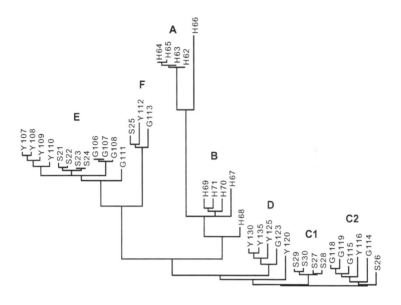

Fig. 3. Tree plot resulting from neighbor joining clustering of forty samples across the K/Pg boundary of the four studied sections

12.5 Acknowledgements

The authors would like to thank Prof. Dr. Hanspeter Luterbacher of the Museum of Paleontology, Barcelona University, Spain, and Dr. David Horne of the Department of Geography, Queen Mary, University of London, UK, for continuous help thoughout the present study and critical reading of the first draft of this paper. We are much indebted to Dr. Oyvind Hammer of the Natural History Museum, University of Oslo, Norway, for explaining some of the techniques used in his PAST software.

References

Dakrory AM (2002) Biostratigraphy, paleoenvironment and tectonic evolution of the Late Cretaceous- Early Paleogene succession on the North African Plate (Sinai, Egypt) and a comparison with some European and Asian sections. Tuebinger Geowissen. Arbeiten (TGA) 64: 263 pp

Hammer O, Harper DAT, Rayan PD (2001) PAST: Palaeontological statistics software package for education and data analysis. Palaeontologia Electronica 4 (1): 9 pp

Keller G (2002) *Gumbelitria*-dominated late Maastrichtian planktic foraminiferal assemblages mimic early Danian in central Egypt. Marine Micropaleont 47 (2002): 71-99

Keller G (2003) Biotic effects of impacts and volcanism. Earth and Planetary Science Letters 215 (2003): 249-264

MacLeod N (in press) End-Cretaceous extinctions. In Selley RC, Cocks, LRM, Plimer IR (eds): Encyclopedia of Geology. Academic Press, London 346 pp

Molina E, Arenillas I, Arz JA (1996) The Cretaceous/Tertiary boundary mass extinction in planktic foraminifera at Agost, Spain. Rev Micropaléont 39 (3): 225-243

Prothero DR (1998) Bringing fossils to life: An introduction to paleobiology. WCB/McGrow-Hill, USA, 560 pp

Said R (1961) Tectonic framework of Egypt, and its influence on the distribution of foraminifera. AAPG Bull 45: 198-218

Twitchett RJ (2006) The palaeoclimatology, palaeoecology and palaeoenvironmental analysis of mass extinction events. Palaeogeogr Palaeoclimat Palaeoecol 232 (2006): 190-213

13 Quaternary extinctions in Southeast Asia

Julien Louys

School of Biological, Earth and Environmental Sciences, University of N.S.W., Sydney, Australia, j.louys@student.unsw.edu.au

13.1 Introduction

The Quaternary extinctions have held the fascination of scientists since the concept of extinction became widely acceptable. In fact, it was the remains of those large beasts, such as the wooly mammoth, who wandered the plains of prehistoric Europe which became one of the integral arguments in the evolutionary debates of the 19th and early 20th centuries (Grayson 1984b). Interest in the megafauna extinction debate has ranged from the purely academic to the highly political, and has been studied by scientists from a range of disciplines including archaeologists, biologists, climatologists, conservationists, geologists, paleontologists, paleoanthropologists, zoologists as well as many others. The extinctions occurred on all continents save Antarctica, and at various times throughout the Pleistocene. Traditionally, the focus of research and debate has been on the Eurasian and North American extinctions but increasingly there has been considerable interest in the Australian extinctions (see, for example, the many references in Reed et al. 2006). The African extinctions have received comparatively less attention, due largely to the fact that they were less severe than any others. Lastly, the South American and, even more so, the Asian Quaternary extinctions have received the least amount of attention. The question of megafauna extinctions has, however, much relevance for today. Increased climatic variability and human-induced environmental degradation occurring throughout the world has resulting in the rapid extinction of many species. An understanding of extinctions, particularly one where we may have played a part, is integral to our ability to mitigate against further loses.

13.2 The Quaternary "megafauna" extinctions

13.2.1 What are megafauna?

The term "megafauna" is one that is not easily defined. Marshall (1984) points out that four different definitions are provided by different authors in Martin and Klein's 1984 *Quaternary Extinctions*, one of the definitive works in the field. He goes on to criticize the use of the term, concluding that it "is a hindrance to understanding the extinction process itself" (Marshall 1984: 796). The use of the term is also examined by Wroe et al. (2004), who demonstrate that its imprecise nature can lead to very different results in extinction studies. In particular, the subjective use of this term can be made to reaffirm any given hypothesis as its injudicious use allows the removal of problematic species and the addition of affirming ones. What then is the solution? Is the term too entrenched in modern work to be discounted? Marshall (1984) suggests that if the term should continue to be employed (and he makes it clear that this is doubtful), then fauna should be divided into "dwarfed megafauna" and "extinct megafauna" to distinguish between those mammals that experienced significant body-size decreases at the end of the Pleistocene, and those that became extinct. However, even if this solution were adopted, then the term "megafauna" still needs to be defined. Furthermore, it is now evident that many species experienced dwarfing prior to extinction (e.g. Guthrie 2003), so these two terms could be used synonymously. Perhaps, as Martin and Klein have done, it is more prudent to refer to this global faunal extinction event as the *Quaternary*, rather than *megafauna* extinctions. This is the approach I have taken, with the understanding that "Quaternary extinctions" refers to the extinctions of the generally large-bodied mammals (i.e. megafauna), which occurred throughout the last 1.8 Ma. In particular, my analyses are restricted to five orders of mainly large-bodied mammals found in Southeast Asia: Carnivora (e.g. tigers), Proboscidea (e.g. elephants), Artiodactyla (e.g. deer), Perrisodactyla (e.g. tapirs) and Primates (e.g. orangutans). For a discussion on the treatment of these orders as megafauna see Louys et al. (2007).

13.2.2 The debate

The debate on the demise of the megafauna is often characterized by two highly polarized points of view: (1) human-induced extinction; and (2) climate-induced extinction. However, the nature and usefulness of these diametrically opposed views has been questioned (e.g. Wroe and Field 2006; Grayson 1984a). It may be that they represent an example of Kuhn's paradigm hypothesis where the core of each view is unassailable (Grayson 1984a). These two points of view are extreme, however, and most megafauna extinction hypotheses fall somewhere in between them. Indeed, there is an increasing awareness that the nature of the extinction will not be in the form of "humans did it" or "climate did it", but rather an understanding that extinctions on each continent were fundamentally different, and that each extinction event will be influenced by different factors, some having a greater impact than others (e.g. Barnosky et al. 2004). A summary of the major points of both paradigms are covered below.

Human-induced extinctions

Epitomized by prehistoric hunters exterminating megafauna upon colonizing new continents, this paradigm covers such theories as blitzkrieg (e.g. Mosimann and Martin 1975; Martin 1984), over-hunting (e.g. Whittington and Dyke 1984), hyper-disease (e.g. MacPhee and Marx 1997), ecological alterations (e.g. Miller et al. 2005; Miller et al. 1999) as well as any other hypothesis that posits extinction were a direct result of human influence. Whether humans were active (as in the former two theories) or passive (as in the latter two theories) in the extinctions is, for the purposes of this review, a moot point. The highlights of the human-induced paradigm are that it provides a ready explanation for:

- why extinctions seem to quickly follow human colonization
- a correlation between Pleistocene extinctions and historical (generally island) extinctions
- why the extinction event was not as severe in Africa relative to the rest of the world; a product of the co-evolution of *Homo* and megafauna such that megafauna were not "naïve" to humans as predators
- why large-sized mammals were more likely to become extinct rather than smaller-sized animals
- why the extinctions affected terrestrial rather than aquatic biotas

However this paradigm fails to account for or fully explaIn

- why dwarfing and extinctions occurred sometimes synchronously, and sometimes asynchronously between taxa
- why some extinctions are not invariably coincident with human colonization

- why megafauna extinctions should even have occurred in Africa
- the presence of non-contemporaneous faunal assemblages during the Pleistocene (i.e. faunal assemblages representing communities which have no modern analogues), and their subsequent disappearance

Several lines of evidence are often cited as proof of human-induced extinction. These include chronology of extinction, the archeological and paleontological records, and computer simulations (Barnosky et al. 2004). Chronology of extinction refers to the apparent synchrony of human colonization and megafaunal extinction. The evidence of close association of these temporal events are strongest for New Zealand (Barnosky et al. 2004). Some authors would argue that a strong case can also be made between these two events for Australia at ~40-50 ka (e.g. Miller et al. 2005, Miller et al. 1999; Roberts et al. 2001). However both timing for megafauna extinction (e.g. Field and Fullagar 2001; Trueman et al. 2006) and human colonization (e.g. Thorne et al. 1999) have been brought into question.

The archaeological evidence most often debated is the presence (or more accurately absence) of kill sites. These sites are the result of hunting, and often butchering, of megafauna by humans. The paucity of these sites has been used by proponents of climatic driven extinction as evidence that overkill is unlikely. Martin (1984), however, has argued that a lack of kill sites would be expected if the extinctions occurred quickly, leaving little time for fossilization to occur. The lack of kill sites therefore does not damage the arguments of proponents of overkill, although as Wroe et al. (2004) point out, nor does it constitute evidence for it. Kill sites are too few and their interpretation in the debate too ambiguous to provide any salient value beyond proving that prehistoric hunters hunted certain taxa.

Globally, the paleontological record is generally used by proponents of climate change, although this type of evidence is equivocal for Australia. Some Australian researches maintain that megafauna are absent in Australian fossil sites dated younger than ~46 ka, the suggested date of colonization by humans (e.g. Roberts et al. 2001). Others (e.g. Wroe and Field 2006) maintain that the date from at least one megafauna bearing site (Cuddie Springs), dated at 30-36 ka, is reliable. If the latter's contention is sustained, this would seriously undermine the potential of a rapid, human-induced extinction of megafauna in Australia. The presence of more reliably dated deposits as well as accepted kill sites in North America and Eurasia means that paleontological datasets are not generally used in arguments advocating human-induced extinctions on those continents.

The last line of evidence to be examined for human induced extinction in this review is that of computer simulations. Models for extinction have been based on simple predator-prey models, where different variables are used to model population numbers. The most basic variables are those of reproduction, mortality and predation rates. Within this basic setup, more complex variables can be added, and different parameters introduced, including rates of migration, effect of dwindling biomass, etc. These are varied in order to determine what ecological parameters might potentially drive animals to extinction via human hunting. There are many criticisms of this line of evidence, the least of which is that most treat all prey as a single species (Barnosky et al. 2004).

Climate-induced extinctions

Proponents of this paradigm propose that the Pleistocene extinctions occurred as a direct result of climate change. Largely those points irreconcilable with the human-induced paradigm as described above are well explained by these theories. Equally, those points well explained by the human-induced paradigm provide difficulties to the climate-induced paradigm. Some lines of evidence are however readily explained by both paradigms, for example, the lack of kill sites, as discussed above.

The Pleistocene was dominated by a succession of glacial and interglacial cycles, with the former associated with more arid climatic conditions, and the latter with more humid ones (see also the more detailed treatment of this issue as it applies to Southeast Asia, below). The most frequent argument used against climate change as a causal factor in the extinction is that similar, and even greater, climatic fluctuations occurred previous to those associated with the extinction event; however these failed to produce significant reductions in megafauna. Another integral argument against climate change as a force in extinction is that it provides no ready explanations of why extinctions seem to follow so closely human colonization of new continents.

A solution to the first criticism follows an ecological threshold model as proposed by a number of authors, in particular Graham (2006). This hypothesis essentially suggests that successive climate changes gradually reduced population numbers of megafauna, until the last major climatic change (usually assumed to be the Last Glacial Maximum (LGM) at ~20 ka), when population levels dropped below a critical level (the threshold). With unsustainable population levels, megafauna eventually became extinct.

The second criticism is less well handled by proponents of climate change, and they cite poor existing chronology of invasions as their primary objection. However, a possible solution to both criticisms was recently advocated by Price and Webb (2006). They cite evidence which suggests that the LGM was in fact associated with the most severe levels of aridity in Australia. This therefore provides a ready explanation for why megafauna became extinct at this time (for Australia at least), and they suggest that the apparent synchronicity of human colonization and extinction of megafauna may in fact be nothing more than a coincidence.

The most important point that the proponents of the climate change paradigm have highlighted is the need to explain disharmonious or non-contemporaneous assemblages of Pleistocene faunas (e.g. Guthrie 1984, Lundelius 1989; Price and Sobbe 2005; Graham et al. 1996). These assemblages lack modern analogues because they exhibit the co-occurrence of taxa which today are found in radically different environments (e.g. Medway 1972, 1977; Cranbrook 2000). The most convincing theory explaining these puzzling co-occurrences suggests that changes in climate during the Pleistocene altered vegetation structures from heterogeneous mosaics to homogenous slates. Proponents of climate change-induced extinctions suggest that the transition from this heterogeneous vegetative structure to the more homogenous conditions we see today resulted in megafauna range reductions, and in many cases, extinctions (e.g. Guthrie 1984). The presence of these disharmonious assemblages is one that human over-hunting proponents find hard to explain, although theories of vegetation change through loss of megaherbivores (e.g. Owen-Smith 1987), or through large scale habitat alteration are notable exceptions.

13.2.3 Towards a reconciliation

Despite the polarity between these two paradigms, there is now an increasing awareness that a single cause is unlikely to be solely responsible for all extinctions, and that both human and climate would have an effect (e.g. Barnosky et al. 2004). The debate has now largely shifted towards how much each factor contributed to the extinction. However, this complex interplay of factors will likely never be completely understood, and Wroe and Field (2006) recently questioned even the validity of this approach. They cite the example of a study of the Caribbean Island lizard (Schoener et al. 2001) where in the wake of a hurricane, one species of lizard was at an elevated risk of extinction from another. As both the hurricane and the presence of the predatory lizard were necessary for extinction, Wroe and Field (2006) argue that trying to determine which of these was the primary factor provides no worthy insight, but in fact only serves to stifle the science. Trying to understand *how* the various factors are likely to have influenced the biota, rather than trying to attach the lion's share of blame to a particular extinction agent represents a more constructive avenue of approach. Only in this way may we gain insights into extinction processes, which may in turn be of benefit in understanding the modern conservation crisis.

13.3 Quaternary Extinctions in Southeast Asia

Although the extinction of large mammals in Southeast Asia has been reported by several authors (e.g. Medway 1972, 1977; Sondaar 1987, Tougard et al. 1996; Cranbrook 2000), it was not until very recently that these extinctions were placed into the global "megafauna extinction" context. Southeast Asia is a complex biogeographical unit, a result of various biotic migrations and endemisms, periods of insularity and unity, orogenies and changes in climate. Therefore, in order to understand the megafauna extinctions in Southeast Asia, one must first understand some of its complex biogeographical history. I will firstly describe those geographical and environmental changes that have affected Southeast Asia's biota. Secondly, I will provide a review of the ecological requirements of those taxa that experienced extreme geographical range reductions and, where possible, those taxa which became extinct. Twenty species of megafauna have an extinction record in more than one Southeast Asian country (Table 1). Mechanisms of extinction will be examined for Southeast Asia within the context of the two extinction paradigms described above. In particular, the possibility of over-hunting of Southeast Asia's megafauna by early hominids, as well as the impacts of climate, vegetation and sea level changes, will be discussed. Finally, a brief examination of the modern extinction crisis in Southeast Asia will be presented.

13.3.1 Geography of Southeast Asia

Southeast Asia, as discussed here, includes Southern China, Myanmar, Thailand, Laos, Cambodia, Vietnam, Malaysia, Borneo and Java. It is generally separated into two distinct provinces, those of Sunda and Indochina (Fig. 1), on the basis of a number of biogeographical differences (e.g. Tougard 2001; Louys et al. 2007). Southeast Asia can also be separated climatically (Chuan 2005) into continental (consisting of Southern China, Myanmar, Thailand, Laos, Cambodia and Vietnam) and insular (consisting of Malaysia, Singapore, Indonesia and the Philippines). The geography of Southeast Asia has been explored in a number of publications (e.g. Hall and Holloway 1998; Gupta 2005), such that this chapter will provide only a brief description of the salient geographical points.

Table 1. Species becoming locally extinct in more than one Southeast Asian country. † indicates globally extinct.; ∞ indicates globally extant; X indicates country of extinction. Adapted from Louys et al. (2007)

Common Name	Species Name	Extinct/Extant	South China	Burma	Laos	Cambodia	Vietnam	Thailand	Malaysia	Java	Borneo
Dubois's antelope	Duboisia santeng	†							X	X	
Yunnan horse	Equus yunnanensis	†	X	X							
Asian gazelle	Gazella sp.	?	X	X							
Giant ape	Gigantopithecus blacki	†	X				X				
Robust macaque	Macaca robustus	†	X		X						
Giant hyena	Pachycrocuta brevirostris	†	X							X	
Chinese rhino	Rhinoceros sinensis	†	X				X				
South Asian rhino	Rhinoceros sivalensis	†		X	X						
Bearded pig	Sus barbatus	†						X		X	
Lydekker's pig	Sus lydekkeri	†			X		X				
Elephant stegodon	Stegodon elephantoides	†	X	X							
Giant tapir	Megatapirus augustus	†	X		X		X				
Indian rhino	Rhinoceros unicornis	?					X	X		X	
Stegodon	Stegodon orientalis	†	X	X			X				
Malayan tapir	Tapirus indicus	?	X			X					X
Serow	Naemorhedus sumatraensis	?	X				X		X		
Giant panda	Ailuropoda melanoleuca	?		X	X		X	X			
Asian spotted hyena	Crocuta crocuta	?	X		X	X	X				
Archaic elephant	Palaeoloxodon namadicus	†	X	X	X		X		X		
Orangutan	Pongo pygmaeus	?	X			X	X	X	X		X

13.3.2 Geological History

The Southeast Asian continental block is largely composed of elements, which had broken off from the southern super-continent Gondwanaland (Gatinsky and Hutchinson 1987; Metcalf 1990, 1996). Much of the modern geographical aspects of Southeast Asia occurred as a result of the fusion between the Sinoburmalaya and Cathaysia plates, in an event known as the Late Triassic Indosinian Orogeny (Hutchinson 2005). More plate collisions followed, including the collision of the Burma plate with Shan highlands in the Cretaceous and the collision between India and

Eurasia in the Eocene (Hutchinson 2005). The latter was considered a "gentle affair" (Ferguson 1993) until the Miocene, where it resulted in the North-South trending mountain ranges of Western Yunnan, Burma and the Malay Peninsula to the east, and the uplift of the Himalayas and the Qinghai-Tibetan plateau to the north (Whitmore 1987).

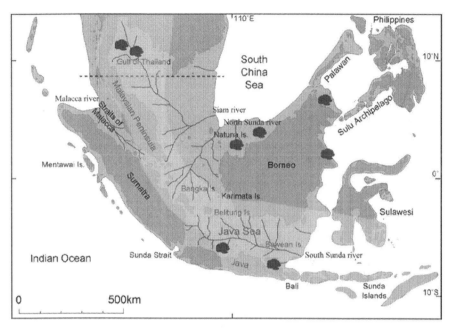

Fig. 1. Southeast Asia at the LGM showing the major rivers and lakes present during that time. Heaney's (1991) proposed savannah corridor is highlighted in a lighter shade. The dotted line represents the division between Indochina (north) and Sunda (south). Adapted after Bird et al. (2005)

The uplift of the Himalayas and the Qinghai-Tibetan plateau during the Miocene significantly altered the established climate, and resulted in the initiation of the Southeast Asian monsoon system (Ferguson 1993). Furthermore, the continued uplift of the Qinghai Plateau during the Pleistocene resulted in a loss of intensity of the summer monsoons during glacial periods, while at the same time increasing the intensity of the winter monsoons (Ferguson 1993). Evidence for these changes in monsoon intensity are also supported by data from ocean cores (e.g. Wang et al. 1999; Gingele et al. 2002). Another, relatively more recent plate collision occurred some 15 Mya between Southeast Asia and Australia, leading to the formation of the Lesser Sunda islands, as well as parts of Sulawesi and the Philippine Archipelago (Hutchinson 1989).

13.3.3 Climate

Three factors have had a major impact on the climate in Southeast Asia over the Quaternary: the position of the Inter-Tropical Convergence Zone (ITCZ), changes in temperature and changes in sea level (Verstappen 1980). The ITCZ and temperature changes will be dealt with here; changes in sea level will be dealt with in a separate section below.

The ITCZ is a low pressure belt situated at or near to the equator, although its position fluctuates throughout the year. In Southeast Asia, its position in December – February passes south of Java and over the northern part of Australia, while in July – September it passes near the Himalayas and the Bay of Bengal, then extends south to near the equator east of New Guinea (Verstappen 1980). Since the maximum amount of precipitation occurs at or near the ITCZ, it has a major impact on Southeast Asia's climate. Verstappen (1980) suggests that the ITCZ had a more southerly position during glacial periods, thereby resulting in an increased dryness in the northern latitudes during these times.

The temperature changes in the Quaternary of Southeast Asia are also related to glacial and interglacial periods. In general the temperature in Southeast Asia has been cooler than is present today. This is supported largely by deep sea cores, which indicate temperatures approximately 3-5°C colder on land, and 2-4°C in the seas (e.g. Chappell et al. 1996; Heaney 1991), however these temperature may have been raised 1-2°C during interglacials (e.g. Verstappen 1997).

Current climate in Southeast Asia varies geographically. Continental Southeast Asia is characterised by greater seasonality, temperature and rainfall extremes, as well as more pronounced dry spells, whereas insular Southeast Asia has a more equable climate (Chuan 2005). The monsoon weather system determines, more so than temperature, the precipitation patterns in Southeast Asia. Two monsoon seasons are present: the summer or southwest monsoon (March – October) and the winter or northeast monsoon (November – April). The summer monsoon brings increased precipitation to continental Southeast Asia, as well as Sumatra and Borneo, while the winter monsoon brings increased precipitation to insular Southeast Asia and Australia. However, distribution of rainfall is also strongly dependent on topographical relief (Chuan 2005).

13.3.4 Vegetation

Vegetation in Southeast Asia is governed by two major factors: water availability and vertical gradient (Corlett 2005). Southeast Asia has been described as the "region of forests climates", which refers to the fact that, until human- generated deforestation within the last few thousand years, the climate of Southeast Asia was such that it supported little else but forest (Corlett 2005). Three major vegetation types for Southeast Asia have been described by Corlett (2005): lowland vegetation (including, among others, tropical rainforests, tropical deciduous forests, savannahs and shrublands), montane vegetation (including montane forests and alpine vegetation) and wetlands.

Tropical rainforests are arguably the most famous of Southeast Asia's vegetation types. As a global percentage, Southeast Asia's rainforests are second only to the neotropics (Heaney 1991; Primack and Corlett 2004). Southeast Asia's tropical rainforests are predominately found in a large block situated in the western part of the Sunda shelf (including Sumatra, the Malay Peninsular and Borneo) (Corlett 2005). Rainforests are also present (or were until recently) in scattered parts of both insular and continental Southeast Asia, including Java and Myanmar (Corlett 2005). The Southeast Asian rainforest is distinguished from other types of rainforests around the world in its dominance of dipterocarps, such that the forest canopy reaches heights of around 30-40 m (Primack and Corlett 2004; Corlett 2005). Tropical rainforests largely occur in regions devoid of regular dry seasons (Primack and Corlett 2004). In regions that experience more extensive dry seasons (i.e. one-four months), the vegetation type changes from tropical rainforest to tropical seasonal forest. These forests are characterised by an increase in the number of deciduous trees, which can at times compose up to 50% of the canopy tree species, although the majority of the forest remains evergreen (Corlett 2005). Tropical deciduous forests occur where rainfall is too low to support a predominantly evergreen forest, generally when the length of the dry season ranges from three to seven months (Corlett 2005). The demarcation between evergreen and deciduous forests is not a gradual one, but usually sharply defined due to the respective forests' tolerance to fire conditions (Corlett 2005).

Other types of vegetation in Southeast Asia include montane vegetation, characterised by shorter, evergreen trees, which are often converted to open woodlands and savannahs by fire during dry seasons; and wetlands, including vegetation especially adapted for soil saturation. Fire regimes seem to be of particular importance in the type of vegetation currently present in Southeast Asia, in particular in the formation of savannahs and shrublands (Corlett 2005). Although these fires can be natural, the last few thousand years have seen an increase in the use of fire by Southeast Asia peoples (Anshari et al. 2001). The effects of humans on the fire regime will be explored in more detail below.

13.3.5 Sea level changes

Changes in sea level have had such an unparalleled influence on Southeast Asia's geography and biota that a detailed treatment of these changes was necessary. Sea level changes occur due to the movement of huge volumes of water from oceans to ice sheets, and back again (Lambeck et al. 2002). These movements are a result of oscillations between glacial and interglacial climatic conditions, and have seen sea levels fall as low as 140 m below present (Lambeck et al. 2002). Regional changes in sea level can be approximated using global sea level fluctuations, which are in turn commonly derived from oxygen isotope studies (e.g. Chappell and Shackleton 1986). Because these oxygen isotopes are derived from marine core drillings, their record can be limited by the depth of the cores. The resulting resolution is therefore much greater for more recent sea level changes but becomes coarser further back in time. Oxygen isotopes are divided into Oxygen Isotope Stages (OIS) (also called Marine Isotope Stages, or MIS), which in turn relate to glacial and interglacial periods. By convention, odd numbered OIS are associated with interglacials, and even numbered OIS associated with glacials (Fig. 2).

During periods of lower sea level, many of the present islands in Southeast Asia were connected to the mainland. Based on modern bathymetric lines, Sumatra would be connected to the Malaysian peninsula if the sea level dropped 30m below present, Borneo at 40m below present, and Java at 50m below present (Bird et al. 2005). These sea-levels are indicated on Figures 2 and 3. However, these values are somewhat simplified: uplift, subsidence and volcanic eruptions occurring throughout the Pleistocene make them an approximation at best (Voris 2000; Bird et al. 2005). Nevertheless, this model serves as a suitable estimate for land connections.

Fluctuations in sea level have resulted in dramatically altered hydrology in the region. Major rivers crossed the exposed Sunda shelf during lower sea-levels (Mollengraaff 1921; Voris 2000). The largest of these is the North Sunda River, which drained northeastern Borneo, the northern parts of the Java Sea and southern Sumatra (Fig. 1). Another large river, the East Sunda River, drained southern Borneo and northern Java, draining in a southeasterly direction. The Siam River is also shown on Figure 1, providing drainage for the eastern Malaysian peninsular as well as the Gulf of Thailand. Many other, smaller tributaries drained other parts of the exposed shelf (Mollengraaff 1921; Bird et al. 2005; Voris 2000). In addition to these rivers, exposed basins on the shelf would have formed large freshwater lakes (Bird et al. 2005).

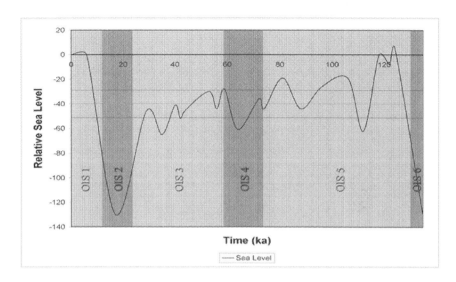

Fig. 2. Global sea-level for the last 130 ka, with associated Oxygen Isotope Stages (OIS). The points at which the major Indonesian islands were connected are indicated by the horizontal lines: from top to bottom Sumatra, Borneo and Java. Relative sea-levels from Chappell & Shackleton (1986); OIS from Martinson et al. (1987)

The existence of a low-lying continental shelf in Southeast Asia resulted in a dramatically different landscape than that found today. Heaney (1991) proposed a 'savannah corridor' running through the middle of Sundaland during the LGM, surrounded on both sides by tropical forests (Fig. 1). Essentially, this savannah corridor would result from a decrease in the surface area of the Sunda Sea, with a resulting loss of evaporation, and hence precipitation in the region (Heaney 1991). This theory has received support from Bird et al. (2005), who examined data from a large number of scientific fields, including geomorphology, biogeography, palynology and vegetation modelling. Although the actual extent of the corridor cannot yet be determined with any level of accuracy, Bird et al. (2005) suggest that a minimal-sized corridor would be 50-150km wide, running between the islands of Bangka, Belitung and Karimata and along the Java Sea through to Java. They also note that little evidence for the existence of such a corridor north of the equator exists.

Although Heaney (1991) proposed the existence of this corridor for the LGM, presumably this corridor would not be restricted to this period, but in all likelihood would also have occurred during other glacial periods. Indeed, evidence for drier conditions in Southeast Asia during the Late

Pliocene and Early Pleistocene exist, including the deposition of thick boulder beds and braided river and alluvial sand sediments (Verstappen 1975, 1997; Batchelor 1979, 1988), as well as palynology records (de Vos et al. 1994; de Vos and Long 2001). The loss of this corridor, resulting from the transition from glacial to interglacial conditions, and its associated effects on the megafauna, is explored below.

13.4 Southeast Asia's megafauna

Unlike North America and Australia, many species of megafauna can still be found living in Southeast Asia. These include such charismatic species as the giant panda (*Ailuropoda melanoleuca*), the orangutan (*Pongo pygmaeus*), the Javan and Sumatran rhinoceroses (*Rhinoceros sondaicus* and *Dicerorhinus sumatraensis*), the Asian elephant (*Elephas maximus*) and the tiger (*Panthera tigris*). Southeast Asia's extinct megafauna is less well known to general audiences, but include several species of stegodons, hyenas and bovids (e.g. *Stegodon orientalis, Pachycrocuta brevirostris* and *Naemorhedus sumatraensis* respectively), as well as the giant tapir (*Megatapirus augustus*) and the giant ape (*Gigantopithecus blacki*). The ecological requirements of those twenty species of megafauna listed in Table 1 are explored below.

13.4.1 Dubois's Antelope

Dubois's antelope (*Duboisia santeng*) is an extinct antelope known from fossil deposits in Java and Malaysia (Louys et al. 2007). At various times this species has been referred to the genera *Bos, Anoa* and *Capra* (see Brongersma (1936) for a discussion on the history of this species), and as such is likely to have shared certain similarities in ecology with these genera. In particular, as a close relative of the extant antelopes, it is likely to have subsisted as a grazer, although this inference should be subject to more detailed study.

13.4.2 Yunnan Horse

The Yunnan horse, (*Equus yunnanensis*), is known only from the northern parts of Southeast Asia during the early Pleistocene (Louys et al. 2007). Like other members of the genus *Equus*, it is likely to have been a grazer, and probably occupied open tracts of grasslands.

13.4.3 Asian Gazelle

Although not identified to the species level, the Asian gazelle (genus *Gazella* sp. indet.) is recorded from Southern China and Burma during the Early Pleistocene (Louys et al. 2007). Although Corbet and Hill (1992) do not record the occurrence of the Asian gazelle in China, the goitered gazelle (*Gazella subgutturosa*) is recorded in the faunal list of Bogdhad Mountain Biosphere Reserve (Information Centre for the Environment 2007), a reserve situated in the far northwest of China (44°00'N/83°00'E). In addition to this species, which is also found in the far west of the Indomalayan region, the Indian gazelle (*Gazella bennettii*) is found in the drier parts of peninsular India (Corbet and Hill 1992). Both extant species are found in open, arid to semi-arid plains (Corbet and Hill 1992).

13.4.4 Giant Ape

An understanding of the ecology of the Asian giant ape (*Gigantopithecus*) has advanced considerably since it was interpreted as "a hunter of large ungulates who apparently dragged his prey back to the cave" (Livingstone 1964: 1284). However, direct interpretations of this fascinating ape's environment have been limited, due largely to the paucity of the fossil material available for study. Ciochon et al.'s (1990) study remains one of a few which deal with this subject matter. Opal phytoliths (opalized remains of plants) found on the molars of *Gigantopithecus* revealed a diet of a variety of fruits and grasses. Furthermore, faunas associated with *Gigantopithecus* suggest a tropical to sub-tropical environment (Kahlke, 1984; Ciochon et al. 1990). It is likely that this ape lived in a similar way to other extant large bodied apes, namely the orangutan and the gorilla, although its large size would in all likelihood have restricted its arboreal capabilities.

13.4.5 Robust Macaque

The robust macaque (*Macaca robustus*) has been found largely in Chinese deposits, but is also known from deposits in Laos (Fromaget 1936). While the palaeoecology of the robust macaque has not been studied in any great detail, an examination of fossil macaques in China indicates that they were all restricted to deciduous broad-leafed forests of sub-tropical to warm temperate zones during the Pleistocene (Pan and Jablonski 1987, Jablonski and Pan 1988). The robust macaque is most closely related to the Japanese macaque (*Macaca fuscata*) and the Formosan rock macaque (*M. cyclopsis*)

(Jablonski and Pan 1988), and as such may share certain ecological similarities with these species.

13.4.6 Giant Hyena

The giant hyena (*Pachycrocuta brevirostris*, formally *Hyaena brevirostris*) is the largest of the true hyenas (Turner and Anton 1996). The postcranial skeleton of the giant hyena indicates that it was not built for running; and although it was only slightly taller at the shoulder than the spotted hyena (*Crocuta crocuta*), it was longer, and its skull was larger and more powerful (Turner and Anton 1996). It is hypothesised as having been an occasional hunter and aggressive scavenger, preying on medium-sized carcasses (Turner and Anton 1996). Its extinction during the Pleistocene has been tied to the world-wide faunal turnover in the felid guild, itself a result of the change in structure of ungulate fauna (Turner and Anton 1996).

13.4.7 Rhinoceroses

Both extant species of Southeast Asian rhinos, the Javan rhinoceros (*Rhinoceros sondaicus*) and the Sumatran rhinoceros (*Dicerorhinus sumatraensis*) are critically endangered. Several species of rhino become extinct in Southeast Asia during the Pleistocene (Louys et al. 2007), including, among others, the extant Indian rhino (*Rhinoceros unicornis*), the South Asian rhino (*R. sivalensis*) and an intermediate form between the Asian and Indian members of *Rhinoceros*, *R. sinensis* (the Chinese rhino). The extant Southeast Asian rhinoceroses are browsers, while the Indian rhino is predominately a grazer (Parr 2003; Groves and Kurt 1972; Laurie et al. 1983). All extant species of rhino wallow, and their preferred habitats include a ready source of water (e.g. river, wetland, streams, etc.) (Parr 2003; Groves and Kurt 1972; Laurie et al. 1983). The Southeast Asian rhinos can be found in tropical and evergreen forests, while the Indian rhino prefers alluvial plains (Parr 2003; Groves and Kurt 1972; Laurie et al. 1983).

13.4.8 Pigs

Two species of pig are listed by Louys et al. (2007) as becoming extinct in more than one country, the bearded pig (*Sus barbatus*) and Lydekker's pig (*Sus lydekkeri*). Pigs subsist in a wide variety of habitats, but prefer those where there is some vegetative cover is available (Nowak 1999). Lekagul and McNeely (1988) noted that they are more common in wet forests, and during the dry season are usually found in riparian environments. They are omnivorous, eating anything from carrion to green vegetation, and can be found in droves of over 100 (Nowak 1999).

13.4.9 Stegodons

The stegodons were a characteristic element of many of the Southeast Asian faunas, although a detailed analysis of its ecology has yet to be completed. In a study of the diet of modern and fossil elephants, Cerling et al. (1999) reported on tooth enamel from a stegodon, *Stegodon* sp., a species from Dhok Pathan, Pakistan, dated at 7.4 Ma old. The $\delta^{13}C$ values obtained for this specimen indicates that C_4 grasses would likely constitute a large part of its diet. Extrapolating the likely diet of Pleistocene stegodons from this one data point is speculative at best. Alternatively, van den Bergh (1999) assumes a more browsing diet for continental (as opposed to insular) stegodons, due to the presence of their low-crowned teeth. This is an area which clearly requires more research. The most common species of stegodon, *Stegodon orientalis*, survived until the early Holocene in Southern China (Tong and Liu 2004).

13.4.10 Malayan Tapir

The Malayan tapir (*Tapirus indicus*) is the only extant species of tapir found outside of South America. It favours well-watered environments with dense forest vegetation (Novarino et al. 2005; Lekagul and McNeely 1988). Although Lekagul and McNeely (1988) suggest they inhabit only primary forests, recent research by Novarino et al. (2005) suggest they actually prefer dense secondary forests. They are generally solitary and most of their activity is tied to rivers and wetlands; they are excellent swimmers and often take refuge in water (Novarino et al. 2005; Lekagul and McNeely 1988). Tapirs are browsers and feed on a variety of aquatic and low-lying vegetation (Nowak 1999; Lekagul and McNeely 1988). In Thailand they are found in tropical evergreen, mixed dipterocarp and mixed deciduous forests (Parr 2003).

13.4.11 Giant Tapir

Despite its description over 80 years ago, little work has been done on the ecology of the giant tapir. Originally described as a subgenus of *Tapirus*, it would appear likely that its habitat and diet were similar to that of the Malayan tapir. Based on its dentition, it was approximately 25% larger than the extant tapir (Tong 2005).

13.4.12 Serow

Members of the genus *Naemorhedus* include the goral and the serow. These goat-like animals are found in a range of forest types, but appear to prefer steep limestone terrain with thick forest (Lekagul and McNeely 1988). They are established climbers, and often shelter in deep forest cover and caves (Parr 2003; Lekagul and McNeely 1988). Like goats they eat a variety of vegetation, but appear to prefer leaves and shoots (Lekagul and McNeely 1988).

13.4.13 Giant Panda

The giant panda, so characteristic of Southern China, was once widespread throughout most of Southeast Asia (Louys et al. 2007). Although subsisting almost exclusively on a diet of bamboo, it has however been observed taking small vertebrate carcasses (Sheng 1999). Due to the heavy dependence of the panda on its preferred food, its prehistoric distribution has been tied to the distribution of temperate bamboo (Tougard et al. 1996). It is currently found in high altitude montane forests, consisting largely of mixed coniferous and broad leafed vegetation (Sheng 1999).

13.4.14 Asian Spotted Hyena

Members of the (extant) spotted hyena, *Crocuta crocuta*, currently restricted to Africa, have also been found in Pleistocene Southeast Asia. In Africa, the spotted hyaena reaches its greatest densities in flat, open country, although it can be found from as diverse habitats as hot and arid areas to dense mist forests (Kruuk 1972).

13.4.15 Archaic elephant

Often put in the same genus as the modern Asian elephant (*Elephas*), *Palaeoloxodon* is actually an extinct, closely related genus (Shoshani and Tassy 2005). However, along with the stegodons and the giant tapir, little research has focused on the specific ecology of this species. Given its high similarity with the modern Asian elephant, however, it is likely to share many habitat and environmental traits. The modern Asian elephant is largely a grazer (Eltringham 1982), and lives in a diverse range of habitats from the forests of Malaysia to the grasslands of Sri Lanka (Eltringham 1982).

13.4.16 Orangutan

The orang-utan is the only extant species of ape (apart from humans) endemic to Southeast Asia. It has a solitary and arboreal lifestyle, although it has been known to descend from the trees. It subsists largely on a diet of fruits, although fungus, leaves, bark honey and insects are sometimes included (Galdikas 1988). Currently restricted to the rainforests of Sumatra and Borneo, like the panda and stegodons, it was much more widely distributed during the Pleistocene (Louys et al. 2007).

13.5 Human overhunting in Southeast Asia?

Synchrony of colonisation and extinctions in Southeast Asia is particularly difficult to establish, and is exacerbated by questions of human evolution and poor chronology of sites (Louys et al. 2007). Unlike North America and Australia, the earliest colonisers in Southeast Asia were not modern humans but *Homo erectus*. Evidence as to whether *Homo erectus* in Asia subsisted predominantly on a scavenging or hunting diet is still equivocal, although current evidence is suggestive that the former scenario is more likely (e.g. Boaz et al. 2004; Schepartz et al. 2005). If this evidence is borne out, it is unlikely that these hominids would have adversely impacted on fauna through over-exploitation.

The earliest records for *Homo sapiens* in Southeast Asia come from Niah Great Caves, Sarawak, dated to approximately 35 ka (Barker et al. 2007). While evidence of hunting is generally accepted for Niah caves, there is no evidence that this was done at an unsustainable rate (Corlett 2007). The predominant prey species appears to have been pigs, followed by primates (Barker at al. 2007; Medway 1977). Bearded pigs are present in all levels of excavation of Niah Caves, suggesting that they were not adversely affected by traditional hunting practices (Corlett 2007). This also suggests, however, that they were not adversely affected by any climate change. Orangutans comprise more than 30% of the non-human primate specimens at Niah, and this high frequency has been explained as a result of human occupants specializing in hunting these apes (Harrison et al. 2006). However given that Borneo is one of only two places in the entirety of Southeast Asia where orangutans still survive, it seems quite unlikely that prehistoric humans could overhunt this species to extinction, as has been suggested for other countries in Southeast Asia (e.g. Harrison et al. 2006). Attributing the extreme geographical decline suffered by the orangutan during the Pleistocene to human overhunting seems premature at this time, given the lack of clear evidence to that fact, coupled with the adverse effects climate change has had on this species (see below).

Furthermore, there is no evidence to suggest that the Pleistocene toolkit of Southeast Asians was adapted to hunting of big game for almost the entirety of the Pleistocene. Stone tool technology in Southeast Asia consisted largely of choppers, and did not share the sophistication of other regions, in particular Europe, until the advent of the Haobinhian during the Late Pleistocene (Reynolds 1990; Corvinus 2004). One explanation for this apparent lack of sophistication suggests the ubiquitous use of bamboo as a substitute for stone tools (e.g. Pope, 1989). However, bamboo appears particularly unsuitable for large game hunting and butchering (West and Louys 2007). In fact, it appears hunting did not become unsustainable in Southeast Asia until the last 2-3000 years, largely as a result of increasing human populations and the advent of modern hunting arms (Sodhi et al. 2004; Corlett 2007).

13.6 Climate change and megafauna

Climate change, and associated fluctuations in sea levels, has dramatically altered the region's vegetation, as discussed above. In particular, the existence of a savannah corridor running through the middle of Sundaland, a concept which seems foreign when we think of Southeast Asia, appears to have been the norm during the Pleistocene. Like other continents, Southeast Asia hosts a number of disharmonious assemblages, and the presence of this corridor provides a ready explanation for these assemblages (Louys et al. 2007; Medway 1972). It is also notable that for the majority of the Pleistocene, Sumatra, Borneo and Java are likely to have been connected to the mainland (Fig. 3).

Of the animals examined above, some prefer open environments over closed. Both the horse and gazelle were largely restricted to the more open north of China throughout the Pleistocene, a result of their preference for more open environments (Jablonski and Whitford 1999). Stable isotope studies show that more open and arid environments were present during the Late Pliocene of central China (Kaakinen et al. 2006), and, given that there existed no physical barrier between the north and south of China during this time, it is likely the southern migration of the horses and gazelles occurred then. The Early Pleistocene of China showed a retreat of steppe-like and temperate zones northwards, with concomitant advances of the tropical and sub-tropical zones (Jablonski and Whitford 1999). The tropical and subtropical zones then subsequently retreated in the Middle to Late Pleistocene, however, steppe-like environments and their associated faunas remained restricted to the north of China due to the physical barrier presented by the rising Qinling Mountains (Ferguson 1993), a barrier which continues to act today (Xie at al. 2004). The horse and gazelle were therefore most likely relicts from a Pliocene southern incursion, becoming eventually extinct as a result of loss of their preferred habitat.

Further south, the Indian rhino became extinct in Thailand, Vietnam and Java in the Middle to Late Pleistocene. Being predominately a grazer, it would conceivably be adversely affected by the loss of the central savannah corridor. Indeed, its current decline in India has been tied largely to loss of grazing land (Laurie et al. 1983). Likewise, grazers and other open-adapted faunas such as Dubois's antelope, perhaps the stegodons, and the archaic elephant *Palaeoloxodon* would also be adversely affected by the loss of grasslands.

The loss of these grazers would detrimentally affect the major scavengers of the region, namely the hyenas. Modern spotted hyenas, although inhabiting diverse habitats, are more common in open grasslands, where carcasses are more easily accessed, and their abundance is directly proportional to the abundance of ungulates (Kruuk 1972).

Many extinct Southeast Asian mammals appear to be intrinsically tied to freshwater sources.

In particular, the rhinoceroses, the Asian hippopotamus (not discussed here, but for an account of this species see Jablonski (2004) and Louys et al. (2007)), the tapir, the giant tapir, and to a lesser extent the pigs, all required ready access to either standing bodies of water or riparian environments. The extensive loss of these sources of water through rises in sea-level, and concomitant changes in the hydrological regime of the region could only have had a negative ecological impact on these taxa. These changes are likely to have contributed, at least in part, to their range reductions, and in some cases, extinctions.

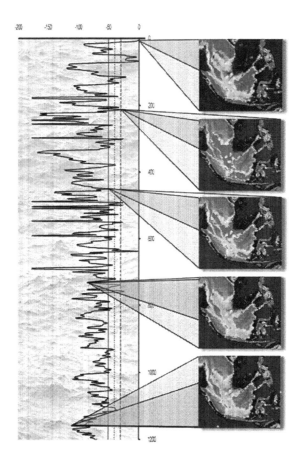

Fig. 3. Global sea-level (x-axis) for the past 1.2 Ma (y-axis). The points at which the major Indonesian islands were connected are indicated by the vertical lines: Sumatra (dashed), Borneo (dotted) and Java (solid). Snapshots of Southeast Asia at selected relative sea-levels are shown. Relative sea-level calculated from Chappell and Shackleton (1986), relative ages calculated from Shackleton and Opdyke (1976), maps of Southeast Asia from the Field Museum of Natural History (2006) (Voris 2000; Sathiamurthy and Voris 2006)

Gigantopithecus and the orang-utan were also likely adversely affected by changes in climate. The ultimate cause of their extinction has been reviewed by a number of researchers (e.g. Jablonski et al. 2000; Jablonski and Whitford 1999). According to their research, it appears likely that these species were driven to extinction or range reduction as a result of the deterioration of their preferred environments during the Middle to Late Pleistocene, in particular the reduction in the tropical and subtropical zones experienced in the north of Southeast Asia.

13.7 The modern extinction crisis

Although the Pleistocene shows relatively little effects of humans on Southeast Asia's biota, the last 10 000 years or so has seen unprecedented levels of anthropogenic pressure; in almost all cases negative. There is general consensus that the massive deforestation which has occurred in earnest since the 1800s as a result of an expanding agricultural economy has been the driving force behind most extinctions (e.g. Sodhi et al. 2004). However there has also been an increasing awareness of the detrimental effects of unsustainable hunting and wildlife trade in the region (e.g. Sodhi et al. 2004; Corlett 2007). Five species of mammals are currently listed as extinct in Southeast Asia (IUCN 2007), with many more critically endangered. It has been estimated that if human-induced deforestation continues unabated, three-quarters of Southeast Asia's original forest cover will be gone by the end of the century (Sodhi et al. 2004), wreaking unknown devastation on its biota. Furthermore, wildlife is being removed from Southeast Asia's tropical forests through hunting and wildlife trade at six times the sustainable rate (Sodhi et al. 2004); such that the density of large mammals even in protected areas are lower than they should be expected (Steinmetz et al. 2006; Corlett 2007).

The timeline of anthropological escalation of environmental destruction can be roughly traced out, and is very much restricted to the Holocene. Traditional Chinese medicine dates back approximately 5000 years, and is associated with the continued reduction of a number of large vertebrates including tigers, bears, rhinos, monkeys and pangolins (Sodhi et al. 2004). Although evidence exists of hunting in the region from the Late Pleistocene (Niah), evidence for unsustainable hunting does not appear until the last 2-3000 years (Corlett 2007). Extensive habitat destruction is also tied to the recent past - the intensity of man-made fires becomes important only within the last 1400 years (Anshari et al. 2001), while deforestation for agriculture, and more recently logging, has been largely restricted to the last 200 years.

While the number of extinct mammals in Southeast Asia is currently relatively small (compared to the rest of the world), this is likely to change dramatically in the near future should current deforestation and wildlife trade practices continue. Even if they do change, however, it is quite likely that the momentum of past activities will have a detrimental impact on the survival of Southeast Asian biota in the future (Brook et al. 2006). But the severity of such potential impacts may be partly mitigated if conservation practices are implemented now. Although this will be difficult due to Southeast Asia's complicated socioeconomic status, it needs to be driven in part by a larger global awareness of this looming disaster. As Dudgeon (2000) notes, tropical Asia does not invoke the same conservation concern as Africa and the neotropics. This is astounding considering that Indonesia, for example, hosts more birds and flowering plants than the whole of Africa (Dudgeon, 2000), and that up to 42% of Southeast Asia's biodiversity could be extinct by 2100 (Sodhi et al. 2004). I certainly agree with Corlett (2007), that unless something is done soon, we will witness the mechanisms of megafauna extinctions in action.

13.8 Summary

Limited evidence suggests that humans contributed to the extinction of Southeast Asia's megafauna until the Holocene. Rather, major changes in vegetation, hydrology and geography, resulting in large part from fluctuations in sea level, have had severe impacts on many large-bodied mammals, in many cases leading to their eventual extinction. Those taxa that did not become extinct often underwent severe restrictions in geographic distribution. Southeast Asia's extant megafauna have not been left unscathed by these Pleistocene environmental throes, and the beginning of the Holocene finds many of them, in a figurative sense, severely battered and just holding on. Enter modern humans, and this precarious situation has been severely tested. In some cases, this test has already failed. However, as Sodhi et al. (2004) state, this is just the tip of the iceberg. It may in fact be already too late to save those species which are critically endangered. And if nothing is changed, many more than these will follow.

13.9 Acknowledgements

Many thanks to Yamila Gurovich, Vera Weisbecker, Pierre Louys, Geoff Mulhearn and especially Gilbert Price for their helpful comments on various drafts of this manuscript, and Guillaume Louys for much needed assistance with the figures.

References

Anshari G, Kershaw AP, van der Kaars S (2001) A Late Pleistocene and Holocene pollen and charcoal record from peat swamp forest, Lake Sentarum Wildlife Reserve, West Kalimantan, Indonesia. Palaeogeogr Palaeoclimat Palaeoecol 171: 213-228

Batchelor BC (1979) Discontinuous rising Late Cainozoic eustatic sea-levels, with special reference to Sundaland, southeast Asia. Geologie en Mijnbouw 58: 1-20

Batchelor DAF (1988) Dating of Malaysian fluvial tin placers. Journal of Southeast Asian Earth Sciences 2: 3-14

Barker G, Barton H, Bird M, Daly P, Datan I, Dykes A, Farr L, Gilbertson D, Harrison B, Hunt C, Higham T, Kealhofer L, Krigbaum J, Lewis H, McLaren S, Paz V, Pike A, Piper P, Pyatt B, Rabett R, Reynolds T, Rose J, Rushworth G, Stephens M, Stringer C, Thompson J, Turney C (2007) The 'human revolution' in lowland tropical Southeast Asia: the antiquity and behaviour of anatomically modern humans at Niah Cave (Sarawak, Borneo). J Hum Evol 52: 243-261

Barnosky A, Koch P, Feranec R, Wing S, Shabel A (2004) Assessing the causes of Late Pleistocene extinctions on the continents. Science 306: 70-75

Bird MI, Taylor D, Hunt C (2005) Palaeoenvironments of insular Southeast Asia during the Last Glacial Period: a savanna corridor in Sundaland? Quaternary Science Reviews 24: 2228-2242

Boaz NT, Ciochon RL, Xu Q, Liu J (2004) Mapping and taphonomic analysis of the *Homo erectus* loci at Locality 1 Zhoukoudian, China. J Hum Evol 46: 519-549

Brongersma LD (1936) Some comments upon H. C. Raven's paper: "Wallace's line and the distribution of Indo-Australian mammals." Arch Néerl zool 2: 240-256

Brook BW, Bradshaw CJA, Koh LP, Sodhi NS (2006) Momentum drives the crash: mass extinction in the tropics. Biotropica 38: 302-305

Cerling TE, Harris JM, Leakey MG (1999) Browsing and grazing in elephants: the isotope record of modern and fossil proboscideans. Oecologia 120: 364-374

Chappell J, Omura A, Esat T, McCulloch M, Pandolfi J, Ota Y, Pillans B (1996) Reconcilliation of late Quaternary sea levels derived from coral terraces at

Huon Peninsula with deep sea oxygen isotope records. Earth Planet Sci Let 141: 227-236

Chappell J, Shackleton NJ (1986) Oxygen isotopes and sea level. Nature 324: 137-140

Chuan GK (2005) The climate of Southeast Asia. In Gupta A (ed) The physical geography of Southeast Asia. Oxford University Press, Oxford pp. 80-93

Ciochon RL, Piperno DR, Thompson RG, (1990) Opal phytoliths found on the teeth of the extinct ape *Gigantopithecus blacki*: implications for paleodietary studies. Proceedings of the National Academy of Science 87: 8120-8124

Corbet GB, Hill JE (1992) The Mammals of the Indomalayan Region. Oxford University Press, New York, 496 pp

Corlett RT (2007) The impact of hunting on the mammalian fauna of tropical Asian forests. Biotropica 39: 292-303

Corlett RT (2005) Vegetation. In Gupta A (ed) The physical geography of Southeast Asia. Oxford University Press, Oxford pp. 105-119.

Corvinus G (2004) *Homo erectus* in East and Southeast Asia, and the questions of the age of the species and its association with stone artifacts, with special attention to hand-axe-like tools. Quaternary International 117: 141-151

Cranbrook (2000) Northern Borneo environments of the past 40,000 years: archaeozoological evidence. The Sarawak Museum Journal 55: 61-109

de Vos J, Sondaar PY, van den Bergh GD, Aziz F (1994) The *Homo* bearing deposits of Java and its ecological context. Courier Forschung-Institut Senckenberg 171: 129-140

de Vos J, Long VT (2001) First settlements: relations between continental and insular Southeast Asia. In Sémah F, Falguères C, Grimaud-Hervé D, Sémah A (eds) Origine des peuplements et chronologie des cultures Paléolithiques dans le Sud-est Asiatique. Semenanjung, Paris, pp. 225-249

Dudgeon D (2000) The ecology of tropical Asian rivers and streams in relation to biodiversity conservation. Annual Review of Ecology and Systematics 30: 239-263

Eltringham SK (1982) Elephants. Blandford Press, Dorset, U.K: 262 pp

Ferguson DK (1993) The impact of Late Cenozoic environmental changes in East Asia on the distribution of terrestrial plants and animals. In Jablonski NG (ed) Evolving landscapes and evolving biotas of East Asia since the Mid-Tertiary. Centre of Asian Studies, Hong Kong, pp. 145-196

Field J, Fullager R (2001) Archaeology and Australian megafauna. Science 294: 7a

Field Museum of Natural History (2006) http://www.fmnh.org/research_collections/ zoology/zoo_sites/seamaps/Accessed 25th May 2007

Fromaget J (1936) Sur la stratigraphie des formations récentes de la Chaine annamitique septentrionale et sur l'existence de l'Homme dans le Quaternaire inférieur de cette partie de l'Indochine. Comptes Rendus de l'Academie des Sciences 203 : 738-741

Galdikas BMF (1988) Orangutan diet, range, and activity at Tanjung Puting, Central Borneo. International Journal of Primatology 9: 1-35

Gatinsky YG, Hutchinson CS (1987) Cathaysia, Gondwanaland and the Palaeotethys in the evolution of continental Southeast Asia. Geological Society of Malaysia Bulletin 20: 179-199

Gingele FX, de Deckker P, Girault A, Guichard F (2002) History of the South Java Current over the past 80ka. Palaeogeogr Palaeoclimat Palaeoecol 183: 247-260

Graham RW (2006) A self-organizing, threshold model for the environmental cause of terminal Pleistocene extinctions in North America. Alcheringa Special Issue 1: 443-444

Graham RW, Lundelius EL, Graham MA, Schroeder EK, Toomey RS, Anderson E, Barnosky AD, Burns JA, Churcher CS, Grayson DK, Guthrie RD, Harington CR, Jefferson GT, Martin LD, McDonald HG, Morlan RE, Semken HAJ, Webb SD, Werdelin L, Wilson MC (1996) Spatial response of mammals to late Quaternary environmental fluctuations. Science 272: 1601-1606

Groves, CP, Kurt F (1972) *Dicerorhinus sumatrensis*. Mammalian Species 21: 1-6

Grayson DK (1984a) Explaining Pleistocene extinctions: thoughts on the structure of a debate. In Martin PS, Klein RG (eds) Quaternary Extinctions: A prehistoric revolution. Tucson, University of Arizona Press, pp 807-823

Grayson DK (1984b) Nineteenth century explanations of Pleistocene extinctions: a review and analysis. In Martin PS, Klein RG (eds) Quaternary Extinctions: A prehistoric revolution. Tucson, University of Arizona Press, pp 5-39

Gupta A (2005) The physical geography of Southeast Asia. Oxford University Press, Oxford, 440 pp

Guthrie RD (2003) Rapid body size decline in Alaskan Pleistocene horses before extinction. Nature 426: 169-171

Guthrie RD (1984) Mosaics, allelochemics, and nutrients: an ecological theory of late Pleistocene megafaunal extinctions. In Martin PS, Klein RG (eds) Quaternary extinctions: A prehistoric revolution. Tucson, University of Arizona Press, pp. 259-298

Hall R, Holloway JD (1998) Biogeography and geological evolution of SE Asia. Backhuys Publishers, Leiden, 417 pp

Harrison T, Krigbaum J, Manser J (2006) Primate biogeography and ecology on the Sunda shelf islands: a paleontological and zooarchaeological perspective. In Lehman SM, Fleagle JG (eds) Primate biogeography: progress and prospects. Springer Science, New York, pp. 331-372

Heaney LR (1991) A synopsis of climatic and vegetational change in Southeast Asia. Climatic Change 19: 53-61

Hope G (2005) The Quaternary in Southeast Asia. In Gupta A (ed) The physical geography of Southeast Asia. Oxford University Press, Oxford pp. 24-37

Hutchinson CS (2005) The geological framework. In Gupta A (ed) The physical geography of Southeast Asia. Oxford University Press, Oxford pp. 3-23

Hutchinson CS (1989) Geological evolution of South-East Asia. Oxford Monographs on Geology and Geophysics, Clarendon Press, Oxford, 376 pp

Information Centre for the Environment (2007)http://www.ice.ucdavis.edu/ bioinventory/bioinventory.html. Accessed 10th May 2007
IUCN (2007) http://www.redlist.org. Accessed 10th May 2007
Jablonski NG (2004) The hippo's tale: how the anatomy and physiology of Late Neogene *Hexaprotodon* shed light on Late Neogene environmental change. Quaternary International 117: 119-123
Jablonski NG, Whitford MJ, Roberts-Smith N, Xu Q (2000) The influence of life history and diet on the distribution of catarrhine primates during the Pleistocene in eastern Asia. J Hum Evol 39: 131-157
Jablonski N, Whitford M (1999) Environmental change during the Quaternary in East Asia and its consequences for mammals. Records of the Western Australian Museum. Supplement No. 57: 307-315
Jablonski NG, Pan Y (1988) The evolution and palaeobiogeography of monkeys in China. In Whyte P, Aigner JS, Jablonski NG, Taylor G, Walker D, Wang PX (eds) The palaeoenvironment of East Asia from the mid-Tertiary, Centre of Asian Studies, University of Hong Kong, pp 849-867
Kaakinen A, Sonninen E, Lunkka JP (2006) Stable isotope record in paleosol carbonates from the Chinese Loess Plateau: implications for late Neogene paleoclimate and paleovegetation. Palaeogeogr Palaeoclimat Palaeoecol 237: 359-369
Kahlke HD (1984) Paleo-environment of Pleistocene *Gigantopithecus blacki* of continental Southeast Asia. International Journal of Primatology 5: 395
Kruuk H (1972) The Spotted Hyena: a study of predation and social behaviour. The University of Chicago Press, Chicago, 335 pp
Lambeck K, Esat TM, Potter E (2002) Links between climate and sea levels for the past three million years. Nature 419: 199-206
Laurie WA, Lang EM, Groves CP (1983) *Rhinoceros unicornis*. Mammalian Species 211: 1-6
Lekagul B, McNeely JA (1988) Mammals of Thailand. Association for the Conservation of Wildlife, Saha Karn Bhaet Co, Thailand, 758 pp
Livingstone FB (1965) Controversy regarding *Gigantopithecus*. American Anthropologist 67: 1283-1284
Louys J, Curnoe D, Tong H (2007) Characteristics of Pleistocene megafauna extinctions in Southeast Asia. Palaeogeogr Palaeoclimat Palaeoecol 243, 152-173
Lundelius EL (1989) The implications of disharmonious assemblages for Pleistocene extinctions. Journal of Archaeological Science 16: 407-407
MacPhee RDE, Marx PA (1997) The 40, 000-year plague: humans, hyperdisease, and first-contact extinctions. In Goodman SM, Patterson BD (eds) Natural Change and Human Impact in Madagascar. Smithsonian Institute Press, pp 169-217
Marshall LG (1984) Who killed Cock Robin? An investigation of the extinction controversy. In Martin PS, Klein RG (eds) Quaternary extinctions. A prehistoric revolution. Tucson, Arizona, The University of Arizona Press, pp 785-806

Martin PS (1984) Prehistoric overkill: The global model. In Martin PS, Klein RG (eds) Quaternary extinctions: a prehistoric revolution. University of Arizona Press, Tucson, pp 354-403

Martinson DG, Pisias NG, Hays JD, Imbrie J, Moore TC, Shackelton NJ (1987) Age dating and the orbital theory of the Ice Ages: development of a high-resolution 0 to 300,000-year chronostratigraphy. Quaternary Research 27: 1-29

Medway (1972) The Quaternary era in Malesia. In Ashton PS, Ashton M (eds) Miscellaneous series; Aberdeen, Scotland, University of Hull; University of Aberdeen pp 63-83

Medway (1977) The Niah excavations and an assessment of the impacts of early man on mammals in Borneo. Asian Perspectives 20: 51-69

Metcalf I (1990) Allochthonous terrane processes in Southeast Asia. London: Royal Society Philosophical Transaction A331: 625-640

Metcalf I (1996) Pre-Cretaceous evolution of SE Asian terranes. In Hall R, Blundell D (eds) Tectonic evolution of Southeast Asia. Geological Society Special Publication 106, London, pp 97-122

Miller GH, Fogel ML, Magee JW, Gagan MK, Clarke SJ, Johnson BJ (2005) Ecosystem collapse in Pleistocene Australia and a human role in megafauna extinction. Science 309: 287-290

Miller GH, Magee JW, Johnson BJ, Fogel ML, Spooner NA, McCulloch MT, Ayliffe LK (1999) Pleistocene extinction of *Genyornis newtoni*: human impact on Australian megafauna. Science 283: 205-208

Mollengraaff GAF (1921) Modern deep-sea research in the East Indian archipelago. Geographical Journal 57: 95-121

Mosimann JE, Martin PS (1975) Simulating overkill by Paleoindians. American Scientist 63: 303-313

Novarino W, Kamilah SN, Nugroho A, Janra MN, Silmi M, Syafri M (2005) Habitat use and density of the Malayan Tapir (*Tapirus indicus*) in the Taratak Forest Reserve, Sumatra, Indonesia. Tapir Conservation 14/2: 28-30

Nowak RM (1999) Walker's mammals of the world. The John Hopkins University Press, London, 1936 pp

Olsen JW, Ciochon RL (1990) A review of evidence for postulated Middle Pleistocene occupations in Viet Nam. J Hum Evol 19: 761-788

Owen-Smith N (1987) Pleistocene extinctions: the pivotal role of megaherbivores. Paleobiology 13: 351-362

Pan Y, Jablonski NG (1987) The age and geographical distribution of fossil cercopithecids in China. Human Evolution 2: 59-69

Parr JWK (2003) Large mammals of Thailand. Sarakadee Press, Bangkok, 203 pp

Pope GG (1989) Bamboo and human evolution. Natural History 10: 49-56

Price GJ, Webb GE (2006) Late Pleistocene sedimentology, taphonomy and megafauna extinction on the Darling Downs, southeastern Queensland. Australian Journal of Earth Sciences 53: 947-970

Price GJ, Sobbe IH (2005) Pleistocene palaeoecology and environmental change on the Darling Downs, Southeastern Queensland, Australia. Memoirs of the Queensland Museum 51: 171-201

Primack R, Corlett R (2005) Tropical Rain Forests: an ecological and biogeographical comparison. Blackwell Publishing, Malden, USA, 319 pp

Reed L, Bourne S, Megirian D, Prideaux G, Young G, Wright A (2006) Proceedings of CAVEPS 2005. Alcheringa Special Issue 1: 475 pp

Reynolds TEG (1990) The Hoabinhian: a review. In Barnes GL (ed) Bibliographical reviews of Far Eastern archaeology. Oxbow Books, Oxford, pp. 1-30

Roberts RG, Flannery T, Ayliffe LK, Yoshida H, Olley JM, Prideaux GJ, Laslett GM, Baynes A, Smith MA, Jones R, Smith BL (2001) New ages for the last Australian megafauna: continent-wide extinction about 46,000 years ago. Science 292: 1888-1892

Sathiamurthy E, Voris HK (2006) Maps of Holocene Sea Level Transgression and Submerged Lakes on the Sunda Shelf. The Natural History Journal of Chulalongkorn University, Supplement 2: 1-43

Schepartz LA, Stoutamire S, Bekken DA (2005) *Stegadon orientalis* from Panxian Dadong, a Middle Pleistocene archaeological site in Guizhou, South China: taphonomy, population structure and evidence for human interaction. Quaternary International 126-28: 271-282

Schoener TW, Spiller DA, Losos JB (2001) Predators increase the risk of catastrophic extinction of prey populations. Nature 412: 183-186

Shackelton NJ, Opdyke ND (1976) Oxygen-isotope and paleomagnetic stratigraphy of Pacific core V28-239 late Pliocene to latest Pleistocene. Geological Society of America Memoir 145: 449-464

Sondaar PY (1987) Pleistocene man and extinctions of island endemics. Mémoires de la Société Géologique de France NS 150: 159-165

Sodhi NS, Koh LP, Brook BW. Ng PKL (2004) Southeast Asian biodiversity: an impending disaster. Trends in Ecology and Evolution 19: 654-660

Sheng H, Ohtaishi N, Lu H (1999) The mammalian of China. China Forestry Publishing House, Beijing, 297 pp

Shoshani J, Tassy P (2005) Advances in proboscidean taxonomy & classification, anatomy & physiology, and ecology & behavior. Quaternary International 126-128: 5-20

Steinmetz R, Chutipong W, Seuaturien N (2006) Collaborating to conserve large mammals in Southeast Asia. Conservation Biology 20: 1391-1401

Thorne A, Grün R, Mortimer G, Spooner NA, McCulloch M, Taylor L, Curnoe D (1999). Australia's oldest human remains: age of the Lake Mungo 3 Skeleton. J Hum Evol 36: 591-612

Tong H (2005) Dental characters of the Quaternary tapirs in China, their significance in classification and phylogenetic assessment. Geobios 38: 139-150

Tong H, Liu J (2004) The Pleistocene-Holocene extinctions of mammals in China. In Dong W (ed) Proceedings of the Ninth Annual Symposium of the Chinese Society of Vertebrate Paleontology. China Ocean Press, Beijing, pp. 111-119 (in Chinese with English abstract)

Tougard C, Chaimanee Y, Suteethorn V, Triamwichanon S, Jaeger JJ (1996) Extension of the geographic distribution of the giant panda (*Ailuropoda*) and search for the reasons of its progressive disappearance in Southeast Asia during the latest Middle Pleistocene. C. R. Acad Sci, Paris 323: 973-979

Tougard C (2001) Biogeography and migration routes of large mammal faunas in South-East Asia during the Late Middle Pleistocene: focus on the fossil and extant faunas from Thailand. Palaeogeogr Palaeoclimat Palaeoecol 168: 337-358

Trueman CNG, Field J, Dortch J, Charles B, Wroe S (2005) Prolonged coexistence of humans and megafauna in Pleistocene Australia. Proceedings of the National Academy of Science 102: 381-8385

Turner A, Antón M (1996) The giant hyaena *Pachycrocuta brevirostris* (Mammalia, Carnivora, Hyaenidae). Geobios 29: 455-468

van den Bergh G (1999) The late Neogene elaphantoid-bearing faunas of Indonesia and their palaeozoogeographic implications. A study of the terrestrial faunal succession of Sulawesi, Flores and Java, including evidence for early hominid dispersal east of Wallace's line. Scripta Geologica 117: 1-419

Verstappen HTh (1975) On palaeo climates and landform development in Malesia. Mod Quart Res SE Asia 1: 3-35

Verstappen HTh (1980) Quaternary climatic changes and natural environments in SE Asia. Geojournal 4: 45-54

Verstappen HTh (1997) The effect of climatic change on Southeast Asian geomorphology. J Quart Sci 12: 413-418

Voris HK (2000) Maps of Pleistocene sea levels in Southeast Asia: shorelines, river systems and time durations. Journal of Biogeography 27: 1153-1167

Wang L, Sarnthein M, Erlenkeuser H, Grimalt J, Grootes P, Heilig S, Ivanova E, Keinast M, Pelejero C, Pflaumann U (1999) East Asian monsoon climate during the Late Pleistocene: high resolution sediment records from the South China Sea. Marine Geology 156: 245-284

West JA, Louys J (2007) Differentiating bamboo from stone tool cut marks in the zooarchaeological record, with a discussion on the use of bamboo knives. Journal of Archaeological Science 34: 512-518

Whitmore TC (1987) Biogeographical evolution of the Malay Archipelago. Clarendon Press, Oxford, 147 pp

Whittington SL, Dyke B (1984) Simulating overkill: Experiments with the Mosimann and Martin model. In Martin PS, Klein RG (eds) Quaternary Extinction: A Prehistoric Revolution. Tucson, University of Arizona Press, pp 451-465

Wroe S, Field J (2006) A review of the evidence for a human role in the extinction of Australian megafauna and an alternative interpretation. Quaternary Science Reviews 25: 2692-2703

Wroe S, Field J, Fullagar R, Jermin LS (2004) Megafaunal extinction in the late Quaternary and the global overkill hypothesis. Alcheringa 28: 291-331

Xie Y, MacKinnon J, Li D (2004) Study on biogeographical divisions of China. Biodiversity and Conservation 13: 1391-1417

14 Current mass extinction

Ashraf M. T. Elewa

Geology Department, Faculty of Science, Minia University, Minia 61519, Egypt, aelewa@link.net

Humanity may either lead to flourishment or to extinction of life and living creatures in our world. Unfortunately, we are going towards the second option. If we look around us, we will surprise for killing ourselves by ourselves. Why we do not try to live in peace with ourselves and with other creatures? Why we hear everyday news on extinction of species of plants and animals alike? Are we going to destroy our life? The answer is in the following lines.

The Wikipedia (the free encyclopedia) defined the Holocene extinction event as a name customarily given to the widespread, ongoing mass extinction of species during the modern Holocene epoch. The extinctions vary from mammoths to dodos, to numerous species in the rainforest dying every year. Because the rate of this extinction event appears to be much more rapid than the "Big Five" mass extinctions, it is also known as the Sixth Extinction. Since 1500 AD, 698 extinctions have been documented by the International Union for Conservation of Nature and Natural Resources. However, since most extinctions are likely to go undocumented, scientists estimate that during the last century, between 20,000 and two million species have become extinct, but the precise total cannot be determined more accurately within the limits of present knowledge. Up to 140,000 species per year (based on Species-area theory) may be the present rate of extinction based upon upper bound estimating. The observed rate of extinction has accelerated dramatically in the last 50 years, to a pace greater than the rate seen during the Big Five (for more information, see the Wikipedia).

The PBS station mentioned that the background level of extinction known from the fossil record is about one species per million species per year, or between 10 and 100 species per year (counting all organisms such as insects, bacteria, and fungi, not just the large vertebrates we are most familiar with). In contrast, estimates based on the rate at which the area of tropical forests is being reduced, and their large numbers of specialized species, are that we may now be losing 27,000 species per year to extinction from those habitats alone. Humanity's main impact on the extinction rate is landscape modification, an impact greatly increased by the burgeoning human population.

The National geographic News clarified that conservationists warn that many birds face the same fate as their prehistoric ancestors, the dinosaurs. But it isn't an asteroid or volcanic eruption that's threatening to finish them off. The culprit, they say, stares at us from the bathroom mirror every day. Humans are singled out, in a recent report, as the cause of what many scientists believe is the biggest mass extinction of animals in 65 million years.

The CNN cued that nearly 2 million species of plants and animals are known to science and experts say 50 times as many may not yet be discovered. Yet most scientists agree that human activity is causing rapid deterioration in biodiversity. Expanding human settlements, logging, mining, agriculture and pollution are destroying ecosystems, upsetting nature's balance and driving many species to extinction.

The Washington Post, in April 21, 1998, cited that a majority of the nation's biologists are convinced that a "mass extinction" of plants and animals is underway that poses a major threat to humans in the next century, yet most Americans are only dimly aware of the problem, a poll says.

The Environment news Service (ENS), in October 4, 1999, affirmed that freshwater species are in at least as much danger as land species. Since 1900, at least 123 freshwater animal species have been recorded as extinct in North America. Common freshwater species, from snails to fish to amphibians, are dying out five times faster than land species, and three times faster than coastal marine mammals.

The CNN, in August 11, 2000, disclosed that reptiles worldwide may be under greater environmental stress than their amphibian cousins, according to a report published in today's issue of the journal BioScience.

The CommonDreams.org website, in August 21, 2001, published that It will take a miracle -- and concerted international effort -- to save the healthy forests left in the world from obliteration, a landmark study from the United Nations says.

The BBC News, in May 21, 2002, argued that almost a quarter of the world's mammals face extinction within 30 years, according to a United Nations report on the state of the global environment.

The UNEP, in September 3, 2002, announced that massive Destruction of Great Ape Habitats Likely Over the Next 30 Years Unless Current Trends Reversed. Less than 10 per cent of the remaining habitat of the great apes of Africa will be left relatively undisturbed by 2030 if road building, mining camps and other infrastructure developments continue at current levels, a new report suggests.

The BBC News, in November 1, 2002, declared that almost half of all plant species could be facing extinction, according to new research by botanists in the United States. The species most at risk live only in small geographic ranges in specific habitats.

The CNN, in May 14, 2003, added that a new global study concludes that 90 percent of all large fishes have disappeared from the world's oceans in the past half century, the devastating result of industrial fishing.

The BBC News, in September 18, 2003, revealed that lion populations have fallen by almost 90% in the past 20 years, leaving the animal close to extinction in Africa, a wildlife expert has warned.

The SFGate.com website, in October 6, 2003, published that decline in oceans' phytoplankton alarms scientists. Experts pondering whether reduction of marine plant life is linked to warming of the seas.

The Science Daily, in July 13, 2004, affirmed that carnivore Species Are Predicted To Be At Increased Extinction Risk From Human Population Growth.

The newsletter of the Ohio Valley Environmental Coalition (Winds of Change Newsletter) Stated, in December 2005, that 10 distinguished scientists including E.O. Wilson and Paul Ehrlich, wrote the U.S. Senate to warn that the Earth is losing species at an unprecedented rate.

The National Science Foundation, in January 11, 2006, disclosed that results of a new study provide the first clear proof that global warming is causing outbreaks of an infectious disease that is wiping out entire frog populations and driving many species to extinction (published in the Jan. 12 issue of the journal Nature).

The BBC News, in January 24, 2006, avowed that the dramatic collapse of orangutan populations has been linked to human activity, new genetic evidence shows.

The Sunday Times, in February 26, 2006, published that researchers have found that carbon dioxide, the gas already blamed for causing global warming, is also raising the acid levels in the sea. The shells of coral and other marine life dissolve in acid. The process is happening so fast that many such species, including coral, crabs, oysters and mussels, may become unable to build and repair their shells and will die out, say the researchers.

There are more and more news on how Man is accelerating extinction of organisms in our world. Therefore, the UN report, dated 3/21/2006, declared that humans are responsible for the worst spate of extinctions since the dinosaurs and must make unprecedented extra efforts to reach a goal of slowing losses by 2010. This means that we need great efforts to save our world from reasons leading to extinction of humankind and other creatures.

Then, what are the important steps to achieve this goal? From my viewpoint, the following steps are to be accomplished:
1. Looking to peace, between peoples, and between them and other creatures, as a strategic goal in the new millennium.
2. Minimizing all kinds of pollution to the least by searching for substitutive products that are not harmful to the environment.
3. Keeping uncommon animals and plants from extinction by making natural protections to them in different areas of the world.
4. Encouraging scientists, especially biologists, to shed more light on this serious problem in the media.
5. Activating the environmental program of the United Nations (UNEP) in the field of protecting plants and animals from loss.
6. Keeping stability of nature by stopping unsafe programs leading to disturbing our environment.
7. Studying the major mass extinctions of the fossil record to know the reasons led to these events.

Of course, there are many other steps that should be carried out to save creatures from extermination, yet the above seven steps might have the priority in this field.

It is surprising, however, that many scientists believe of multiple causes for the five big mass extinctions in the fossil record, but all agree that human activity is the most single cause of deteriorating biodiversity in the recent life!!

15 Current insect extinctions

Panos V. Petrakis

National Agricultural Research Foundation, Institute of Mediterranean Forest Ecosystem Research, Lab. of Entomology, Terma Alkmanos, 1152 Ilissia, Athens, Greece, pvpetrakis@fria.gr

15.1 Introduction

The question 'how many species inhabit the earth?' is not just the result of the inherent property of human mind for continuous intellectual search. It is the urgent need to know the biodiversity of our planet, to predict its fate and increase our understanding of various biological species, their mutual dependencies and their abiotic requirements.

Insects constitute the dominant taxonomic group of the earth. Many authorities summing both the taxonomically described and undescribed species, estimate the total to be from 1.8 to 50 millions (May 1988, 1990; Stork 1993). In an attempt to assess the ratio of the described to undescribed living taxa Hammond (1990, 1992) showed that insects are the richest taxonomic group in both categories. Even the described species surpass all the other groups of organisms in all five kingdoms.

In this respect it is expected that among the organisms that are going to become extinct in the next few decades insects will constitute a large proportion (Stork 1993). Pimm and Raven (2000) estimate that among living species one-tenth may become extinct in the next fifty years. This is expected because of extensive habitat loss, which in a wider sense includes habitat fragmentation and inbreeding suppression among the remaining populations. Despite the fact that insects, especially the strict specialists, are among the first organisms liable to become extinct our knowledge on their taxonomy, distribution and ecological role is very limited. What we need is time and money, for the documentation of insect extinctions (Dunn 2005). In this context the monitoring of insect extinctions is different from that of mammals and birds. Dunn (2005) states that the IUCN Red List (2002) records 129 bird species extinctions within the last 600 years, which constitute 1.3% of all bird species. By analogy we may expect through the world some 44,000 of the estimated total of 3.4 million insect species to become extinct. However, only 70 insect species extinctions have been documented so far. This great discrepancy is largely the result of our limited study of insects. If we accept the arguments of other authors in biodiversity calculus and the estimation of insect species this discrepancy becomes even larger. The "guestimates" of world insect species derived by Stork (1993) is 5.3 million species. Gaston (1992) estimates range between 6.88 and 8.75 million species. Hodkinson and Casson (1991) extensively collected tropical *Hemiptera* species sampled with all known devices from Sulawesi, Borneo estimated the number of insect species of the world to be 3.0 – 4.0 millions. On the same line of arguments by accepting the working figure of 10,000 beetle species in the collection area and taking species accumulation curves the estimates rise to 5.0 – 6.0 million species. Other much higher numbers, though not so convincing are provided by Erwin (1992) who estimated the world insect fauna as 30.0 million, while on the basis of distribution models relating species abundance and body size of various taxa inferred global estimates of insect fauna between 10.0 – 50.0 million (May 1988, 1990).

Surprisingly, the number of studies predicting the richness of the world insect fauna is not balanced with studies predicting the number of insect species to be affected by modern extinctions either on a global or a regional scale. One explanation lies probably in the fact that science is dominated by western schools of thought and associated cultures in which insects and arthropods in general are primarily considered as pests that should be controlled (Kim 1993a, b). An immediate consequence is the exclusion of insects from almost all conservation plans, except possibly those for butterflies and some dragonflies. In the current extinction event, which is induced by humans, the need to separate background extinctions from exceptional extinctions is urgent since it leads to the quantification of local extinctions and the organization of a global network of databases and experts. It is well known to perceptive field entomologists that many insect populations are disappearing from their area of collection and that other species may invade the area establish new populations or even a network of metapopulations depending on the fragmentation status of the habitat. Nevertheless, it is not known if local extinctions are part of a global mass extinction or they are merely above-background events. Paleontologists used to face the problem by interpreting the scarcity or disappearance of insect families as extinctions in excavation layers corresponding to geologic epochs (Labandeira 2005), a fact that reflects the efforts and the cost rather than scientific reality.

In this chapter I have attempted to present the extent of current insect extinctions and relate them to possible taxonomic bias. Also, causes of extinctions and similarities with previously recorded mass extinctions are discussed. Current theories about colonization of an area by insects and the applicability of these to current insect extinctions and their detection is examined. The necessity for extensive taxonomic studies is obvious and the ways that this can be achieved and function in the estimation of species loss are presented.

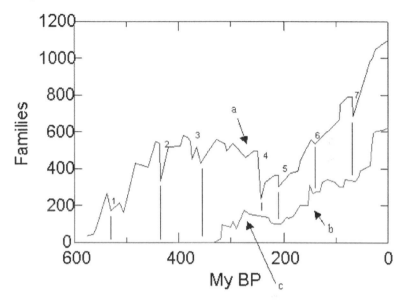

Fig. 1. Major mass extinctions of the past or smaller biotic crises. *X* axis is My BP (= million years before present) and *y* axis is the number of families found in the fossil record. The figure should be consulted in conjunction with Table 1. 1, 2 and 3: early Cambrian, end Ordovician and end Devonian extinctions; all three extinctions happened before the emergence of insects; 4 = end Permian P-T extinction; 5 = mid Triassic extinction; 6 = biotic crisis of mid Cretaceous while 7 = end Cretaceous – early Tertiary K-T extinction. The vertical bars connect the upper all –predominantly marine– animals curve and lower insect diversity curve. (Data and curves are derived from Erwin et al. 1987, Benton 1990, Maxwell 1989 (especially for insects), and also Labandeira 2005 and Labandeira and Sepkoski 1993). **(a)** is the middle Pennsylvanian extinction, **(b)** is the late Jurassic extinction, and **(c)** the extinction indicated by arrow **a** is not followed by origination (see text for explanation)

15.2 Insect mass extinctions in the past

It is known that insects are very different from other organisms at least in the distinction of background and mass extinction. The difference lies in the fact that insects are always dependent on other organisms and usually are important parts of local ecological communities. Labandeira and

Philips (1996) record the late Carboniferous (Middle-Late Pennsylvanian stage – Fig. 1) extinction and consider that this is the first major insect extinction. At the same time an extinction of extensive wetland communities is observed in holarctic areas. The plants replacing the herbaceous forms had mainly arborescent life forms and harboured another insect fauna comprising exophytic and endophytic herbivorous insects that replaced the rich insect fauna of detritivorous taxa and their natural enemies. Typically this type of faunal turnover is seen as extinction in the fossil record.

The second major extinction is that of the end Permian (P-T). According to Erwin (1998) this extinction is very complex and very long in duration – it lasted 1 My BP. It is considered the biggest mass extinction since c. 70 % of terrestrial families and 51% of marine families and 82% of genera disappeared. Benton (1990) states that the major cause was the unification of all continents into one called Pangaea. This triggered a series of events like the extinction of coastal ecosystems in lowland landscapes. The lowland ecosystems disappeared together with all lowland and coastal taxa. This process is similar to the collision of large icebergs to the Antarctic land, which raises extensive barriers to all organisms that hunt fishes in the shallow sea close to the cost and breed in the interior of the continent. Emperor penguins are such animals. By analogy, though much extended in time and area this has occurred in the end Permian extinction. Many explanations and arguments of the gradualist school of thought hold in this extinction. These explanations are coming from vertebrate palaeontologists who put the cooling of the end Permian at a prime (Benton 1990). It seems that species adapted to warm climates did not evolve among homeotherms. For insects, the loss is obvious in this explanation and this is the cause of the difference of modern insect biodiversity between the tropics and the temperate regions.

All authors agree that this large extinction affected families that existed an extended geologic time before (Labandeira 2005). The same author explains this in terms of the Signor-Lipps effect. Signor and Lipps (1982) studying marine microorganisms stated that geological strata poor in fossils and strata rich in fossils are separated in a stepwise fashion that is, fossils corresponding to a certain mass extinction are found in earlier geologic intervals like the steps of a stair, blurring thus the picture of a single extinction event by splitting it into many events. In this explanation, the actual mass extinction took place in the mid-last Permian and lasted at least three stages and not only the last Permian stage.

Another explanation emanates from the catastrophe induced by the impact of an asteroid or comet invoked for the K-T mass extinction and supported by the findings of various workers. Although the catastrophist scenario cannot be excluded, it is very unlike in the case of end-Permian extinction of organisms. This primarily is due to the long duration of fossil findings of a species and the fact that survivors and non-survivors cannot be readily allocated to ecological categories. The same holds for the more recent great extinction, namely K-T. Moreover, some heavily affected groups in the P-T extinction and replaced by other communities are not affected in K-T extinction event (VanValen 1984a, b). Evidently, these groups are not replaced in this extinction. This non-replacement was interpreted as lack of evolution in the short time interval of extinction, which comprises one argument of catastrophists. This is the case the communities of detritivorous animals, which remained little affected in K-T extinction while free-swimming forms (such as plankton, ammonites, belemnites) together with filterfeeders (such as corals, bryozoans, crinoids) were also heavily affected and became extinct.

Table 1. Geological eras, periods, epochs and stages

Phanerozoic eon	Cenozoic era			
		Quaternary period		1.6My - 0
			Holocene epoch	10Ky - 0
			Pleistocene epoch	1.6 My - 10Ky
		Tertiary period		66.4 – 1.6My
		Neogene	Pliocene epoch	5.3 – 1.6My
			Miocene epoch	23.7 – 5.3My
		Palaeogene	Oligocene epoch	38.6 – 23.7My
			Eocene epoch	57.8 – 38.6My
			Palaeocene epoch	66.4 – 57.8My
	Mesozoic era			
		Cretaceous period		144 – 66.4My
		Jurassic period		208 – 144My
		Triassic period		245 – 208My
	Palaeozoic era			
		Permian period		286 – 245My
		Carboniferous period		360 – 286My
		Devonian period		408 – 360My
		Silurian period		438 – 408My
		Ordovician period		505 – 438My
		Cambrian period		570 – 505My
Precamb	Proterozoic era			2500 – 570My
	Archaean era			3800 – 2500My
	Hadean era			4500 – 3800My

Although Van Valen (1984a, b) is very skeptic about the asteroid impact theory as the cause of the K-T extinction and this chapter is not the place to discuss the *pros* and *cons* of extinction theories, it became evident that insects are affected by extinctions indirectly through the disruption of ecosystem processes. In this respect, mass extinctions of the past are not of direct relevance to the current insect extinction. This is due to many reasons. First, the causes of past extinctions are not anthropogenic as is the case of current extinction. Second, the biodiversity measured in past extinctions is mainly at the family level and the fossil records of genera and species is very rare and in many cases illusive. Third, the fossil record for insects is scant, poor and by no means comparable to that of vertebrates. It is believed that fossil insects cannot provide the same variety and depth of the answers provided by vertebrates. Fourth, the extinctions are not always followed by originations of insects (Fig. 1, arrow c) since insects are evolved towards many directions not always related to the factors that permitted their survival from past extinctions (Jablonski 2001) (see also Fig. 1 the branch following the extinction 7 in the insect biodiversity curve). The situation in these interpretations is not clear and the observed originations have actually occurred in previous geological periods. According to Labandeira (2005) the Paleogene (Eocene epoch) extreme originations of insects (Fig. 1, Table 1) have actually occurred in the Cretaceous. This is the effect of the paucity of originations following the K-T extinction (*ibid.*). Fifth, while extinctions are evidenced from common taxa that cease to appear in excavations, in insects the fossilized taxa are not representative of the faunas since many records are coming from individuals trapped in amber. These specimens reflect the living / feeding preference of an insect rather than the prevailing ecological conditions through the most abundant species. Moreover, the survival of an organism from a major extinction does not guarantee that it will be successful and numerous after the conditions leading to the extinction (Jablonski 2002). Whatever the cause of the extinction, the lack of post extinction diversification is surprising given that the majority of taxa experienced population bottlenecks, which are proved evolution promoters.

15.3 Types of current insect extinctions

While it is accepted that current insect extinction are basically a human induced process various authors emphasize individual causative factors. These isolated factors are usually global warming (Thomas et al. 2004) elevated carbon dioxide (Penuelas and Estiarte, 1998), co-extinctions (Dunn 2005; Koh et al. 2004), habitat Zschokke et al. 2000; (Tscharntke and Kruess 1999; Steffan-Dewenter and Tscharntke 2002; Londré and Schnitzer 2006) and habitat loss (Seabloom et al. 2002; Pimm and Raven 2000) or the variously induced decline in ecosystem services (Seastedt and Crossley 1984; Wilson 1987; Dobson et al. 2006; Foster et al. 2006). It is well known that in some regions people are co-responsible for many process either acting towards decreasing or increasing biodiversity (McKinney 2002).

On the other hand there are many cases that human actions, such as introduction of organisms for classical biological control purposes, caused a severe decline in native insects. Boettner et al. (2000) refer that 80% of the larvae of saturniids (Lepidoptera: Saturniidae) are infected by flies introduced in their habitats as biocontrol agents for other insects. It is amazing that in many countries there are no common tests that must be done on introduced biocontrol parasitic agents despite the fact that many cases of local extinctions have been reported.

To know the decrease or the extinction of insect species from certain places one has to know first the bonds with the ecosystems and the role the insects play in them. In this sense the extinction of insects is not separable from the extinctions of other organisms though in this case we know very little about the former (McKinney 1999; Dunn 2005; Samways 1996, 2006). However, all extinctions are amenable to the same analysis common to all organisms. Samways (2006) quoting the works of Mawdsley and Stork (1995) and McKinney (1999), states that the geographically restricted (endemic) insect species are more likely to become extinct though some recent extinctions. This occurred in geographically widespread species like the locust *Melanoplus spretus* (Orthoptera). According to Pimm and Raven (2000) the destruction of many habitats that are colonized by organisms must be converted to species loss. This can only be done through the theories relating the species to the area they colonize. In this respect the predictions of island ecology must be applied and the probabilities of extinction of species must be estimated in time by taking into account all the other factors further to the area of the "island" *per se* (Gorman 1979).

A problem at this point is that all the predictions of the theory of island ecology are referring to communities near or at equilibrium (Krebs 1983). As a rule, extinctions take communities away from equilibrium. There is no way to analyze such communities with mathematical methods unless doing approximations and simplifying assumptions. Indeed, all natural communities and ecosystems are not equilibrium entities and all theories on them are very basic approximations (May 1975; Yodzis 1989).

15.3.1 Lessons from island ecology

Perhaps the first attempt to explain the extinctions of species from islands was done by Ricklefs and Cox (1972; 1978) who investigated the extinction of some bird species from the Lesser Antilles archipelago. The explanation involved the principle of 'taxon cycle', a term which parallels the life of a taxon with the life of an individual. The term was coined by Wilson (1961; see also Ricklefs and Birmingham 2002 for a more recent account of the concept) and it roughly describes the initial spread of a taxon to the islands of an archipelago and the subsequent decline of its populations from many islands. The process ends in the final extinction of the species from the archipelago. Gorman (1979) accepts four stages of this cycle and depicts the case of birds living on the volcanic islands of the Fiji archipelago. In Fig. 2 are shown five habitat types and their use by bird species measured by the value, which is the percentage of a score. Habitat specialists occurring in only one habitat are scored 1 or 100%. Species recorded in two habitats take the score 50%, in three habitats 33% and so on. The average of all species in each habitat in each stage of the cycle is the value corresponding to the species assemblage. The situation revealed in Figure 2 is that as the *taxon cycle* proceeds the species are restricted to two habitats (lowland and montane rainforests) while at the second stage they were evenly distributed to all habitat types. Birds and flying insects are especially suitable to make rapid movements across all habitats by migration while other species colonize the habitats in a much slower and stochastic way (Petrakis and Legakis 2005). These last species are obstacles to the colonization of other habitats. They compete with any new species arriving in the habitat increasing its extinction risk.

The problem of what provides the energy for the taxon cycle to run is crucial in that it describes a basic situation common to all geological periods and geographic areas. The widespread distribution of colonizers at the first stages of the cycle is based in the restricted competitive ability of the resident of the habitat and the finding ability of potential natural enemies. At later stages the competitive ability of the residents and the efficiency of natural enemies are improved and for this they cause the extinction of the species from many habitats.

This situation is shown also in Fig. 3, as a combination of the island equilibrium theory and the resident model. Islands are colonized at a certain rate in each time while the extinction curve is raised in an exponential way being zero at some time at the beginning of the colonization. This is due to the increase of extinction risk when species content increases.

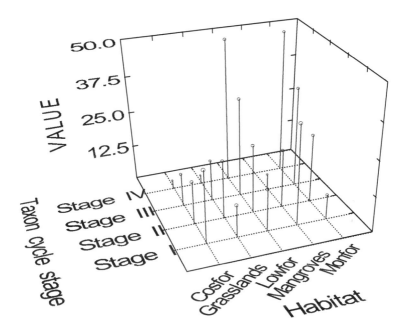

Fig. 2. Habitat use of bird species in habitats of Fiji islands. Birds are given in the different stages of '*taxon cycle*'. Cosfor = coastal forests, Lowfor = Lowland forests, Monfor = Montane forests (explanation in the text)

It is usually accepted that the size of the island affects the extinction rate, which for small islands becomes abrupt. The remoteness of the island usually affects the immigration curve, which for near islands is abrupt and starts from large rates. This is because far islands are reached by mainland organisms with difficulty (Diamond and May 1981; Petrakis 1992). Even in very common colonization events as those of butterflies the species can re-colonize islands or small fragments of suitable biotopes if the source of colonists is close enough to the site. In this way many local extinctions are avoided, and the local population network is maintained. This is the main argument for the design of many small natural reserves that can act as sources of organisms or stepping stones for migration (Dempster 1991; Petrakis 1992; Petrakis and Legakis 2005)

When Pimm and Raven (2000) reported that humans destroy the tropic humid forests suggested that the species loss because of habitat destruction was moderate. While half of the humid forests have been destroyed in this way the species loss was only 15%. This is due to the shape of the extinction curve in figure 3, which is initially zero and then its slope gradually increases. If the destruction of rain forests will continue at the rate estimated now, they predict that in fifty years from now the species loss will peak and then it will start to drop. If we protect efficiently the areas with exceptionally high biodiversity (= biodiversity hot spots) then the species loss will be smaller but it will continue in the same way.

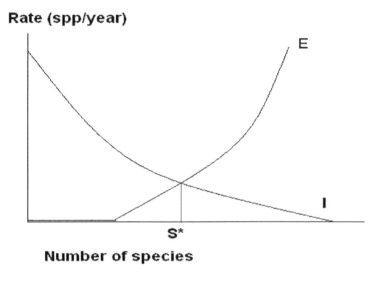

Fig. 3. The combination of the equilibrium model of MacArthur and Wilson (1967) with the resident model of Diamond and May (1981). E= extinction curve, I=immigration curve, S*=equilibrial number of species

15.3.2 Insect extinctions induced by changes in succession status

There are many sites where the probability of extinction is very high. In these places species are lost at a constant rate and local communities very soon become poor (Colinvaux 1973). The widespread farming of the earth surface causes many species extinctions not only through the species area effect. According to many authors plough is a contributing factor more important even from thermonuclear weapons. It keeps communities at an early stage of the ecological succession. At this stage the plant production is high and involves pioneer species properly armed with anti-herbivore defences (Strong et al. 1984). Only a few animals, especially insects, are able to consume this plant biomass, which is consistent with the low biodiversity of pioneer fauna. A well-known animal perpetuates this condition by applying plough once or twice a year to these fields in order to sustain the high plant production for its own benefit. Similarly, in a much smaller scale, other opportunistic species exploit this high plant production. The house sparrows (*Passer domesticus* and *Passer hispaniolensis*) and the robins (*Erithacus rubecula*) near human establishments illustrates this situations (Colinvaux 1973) while other insectivorous birds like swallows exploit the high densities attained by pioneer insect species. These bird presences together with the presences of some rodents like *Rattus norvegicus* do not guarantee or indicate that local biodiversities are kept at high levels similar to those of advanced succession stages. Actually all these species are efficient competitors or natural enemies of insects. Only a few parasitic insects and other arthropods of the vertebrates are maintained but these species by no means compensate for the insect and other species loss.

Another problem that emerges from the maintenance of an early succession stage is the lack of natural disturbances. Many bark beetles prefer stressed or weakened trees while other insects prefer dead wood from fallen trees and branches, commonly termed 'saproxylic insects' (Grove 2002). These resources are characteristic of more advanced succession stages. The lack of occupants in the detritivorous niches causes the accumulation of biomass in the ecosystem. Natural systems have many checkpoints for the control and direction of ecosystem processes. To cope with the inefficiency of the communities of decomposers, natural ecosystems utilize fire to remove the accumulation of organic matter (Trabaud 1991). In addition, fires facilitate the invasion of exotic species superficially increasing in this way the local biodiversity. The net effect of fires is the extinction of indigenous species with a simultaneous establishment of non-native invasive species (McKinney 2002). The fact that the frequency and intensity of fires is positively correlated with human activities is an indicator of the situation that fires are more frequent in populated areas receiving anthropogenic influence. This blurs the cause of fires, which according to political authorities is based on local human interests while according to others fires can be spontaneous caused by the accumulation of dead organic matter. The example of an Australian problem is referred in Morris et al. (1991) where native decomposers – mainly dung beetles– could not metabolize non-marsupial dung. This guided Australian authorities to import dung beetles from Africa in order to decompose the organic matter from cowpats accumulated in the pastures.

In addition, the fire destroys and removes wood before it is lost to decay. The same happens to all managed forests (Grove 2002). The factor that supports a rich saproxylic insect fauna is removed in this way and the mature habitat, rich in various insect assemblages disappears together with the associated insects (Ranius 2002; Ranius et al. 2005; Jonsson et al.

2005; Spies et al. 2006). Altered management practices are not as effective as the protection of forested areas in which a few large preserves are more effective in conserving the biodiversity of dead wood arthropods than many small ones (Ranius and Kindvall 2006).

15.3.3 Insect extinctions from climate change

It is well known that the climate of the earth becomes warmer (Parmesan 1996; Schmitt 2003) and this is going to continue for the next 50-100 years (Parmesan et al. 1999). This situation triggered various associated processes such as the decline of plants, which are important hosts for herbivorous insects (Barber et al. 2000; Scott et al. 2002; Petrakis and Legakis 2005). In this sense global warming cannot be separated from other sources of extinction such as habitat loss and co-extinctions. The loss of suitable habitats is a factor intrinsic to any plant species extinction (Schmitt 2003) while adults are affected by more extensive losses of plant individuals (Cabeza 2003).

To examine in detail the effect of climate change on insects, one safe approach is to examine museum collections of insects in order to trace the various shifts that occurred in the past in the boundaries of the geographical distributions and/or phenological shifts and voltinism (Graham et al. 2004). This approach requires well-organized museums with extensive use of informatics. However, for many regions of the earth museums are not well organized and this kind of information cannot contribute to a better understanding (Petrakis and Legakis 2005).

Many authors have shown that evolutionary response of herbivorous insects to the phenological change of their host plant -as it is expected from global warming (Claridge and Gillham 1992). The response of insects to global warming can result in a reshuffling of local insect assemblages and change in an unpredictable way the insect-plant associations (Jensen 2004). It is believed that insect species predisposed for migratory behavior will be at an advantage if global warming affects their geographical range.

The polarward shift of the northern boundary (= PSNB) of forty butterfly species (Parmesan et al. 1999) is usually considered as a response to the temperature rise and Root et al. (2003) have included this work among those which give support to the ongoing global warming from the side of insects. This finding and the paper of Root et al. (2003) were severely criticized by Idso et al. (2003). The criticism states that the fact that the southern boundary remains stable refutes the hypothesis of elevated CO_2 induced global warming extinction since this can be interpreted as a range expansion. The same authors believe that the observed homogeneity within the geographic range of the Edith's checkerspot butterfly *Euphydryas editha* contradicts the (= PSNB) (Parmesan 1996). Many viable populations exist in all parts of the range. According to Idso et al. (2003) this indicates that the shift is not the result of global warming. In addition they predict that the most parsimonious in energetic terms movement would be an elevation shift, which Parmesan (1996) did not find. Also, Parmesan does not evidence the possibility of increased extinction rates at lower elevations while there is no any simultaneous escape shift to higher elevations.

At the same lines it has been observed by many other authors that several insects have expanded their ranges. This expansion, as a rule, is done with a stable southern boundary and this is interpreted as a proof of non-increasing extinction rates. Hill et al. (1999; 2001) have found that the speckled wood butterfly *Pararge aegeria* (Satyridae) has expanded its geographical range since the contraction at the end of the nineteenth century. Idso et al. (2003) interpret this as a typical case of a butterfly of the Northern Hemisphere, which finds suitable northern biotopes and is spreading over them by creating viable populations in a very slow process, which is somewhat different from the way that that many authors envisage global warming.

Another way to see range expansion induced by climate warming is the increase in resource availability. Beerling and Woodward (1996) state that since the correlation of stomatal density with the ambient concentration of carbon dioxide is weak, it is unlikely that we can found past concentrations by examining the stomatal densities of extinct and extant plants. In effect, it is not clear how the elevated CO_2 level can be reflected to increased plant resources and how this can trigger an increase in population densities (Zurlini et al. 2002; for a discussion see Petrakis and Legakis 2005). In a study of the floristic and vegetation resources at several mountain summits in the context of the GLORIA multi-summit program Kazakis et al. (2006) have not observed elevation shifts of plants at Lefka Ori, Chania, Crete. These authors made only predictions of what is likely to happen in a scenario of temperature rise. More precisely they predict that only southern slopes are likely to be colonized by lowland occupying plants while northern slopes are more resistant to such colonization. This is the case of the endemic *Nepeta sphaciotica* at the northern slopes of Svourichti summit, Lefka Ori, which is expected to face some resistance. The richness of plants at higher altitudes is expected to increase with the spread of thermophilous plants. In this scenario of plant increase the insects are predicted to spread. A case of the attack of the pine processionary caterpillar *Thaumetopoea pityocampa* (Lepidoptera, Thaumetopoeidae) to the boreal pine relict species *Pinus sylvatica* ssp *nevadensis* has been already documented by Hodar et al. (2003). The authors of this work state that the increase of winter temperature causes increased survival of the insect and reduced regenerative capacity of the relict Mediterranean Scots pine. These events cause a gradual elimination of the pine, which can be reversed if the structure of the habitat is properly managed. This is an example of an expansion in the elevation range of an insect species. However, many endemic plants are expected to become extinct because of the spread of thermophilous species, which will be competitively superior. Fortunately, these endemics have no insect species specializing on them; at least no report of such a specialization has been published so far. This situation is expected since most Cretan endemics are palaeoendemics (Greuter 1972; Valentine 1972) and they survived because they evolved very efficient anti-herbivore defenses (Petrakis et al. 2004) against which the host did not evolve any counter-adaptation. In this respect there will be a few co-extinctions of insects as a result of the extinction of their host.

Insect extinctions can be caused by other routes such as the social behavior of many caterpillars. For instance it is known that many ectotherm animals such as caterpillars rely on sociality to increase the temperature of their bodies by the collective production of metabolic heat. In case the caterpillar is alone the heat production is not enough to increase the body temperature of the insect above ambient temperature. Such a case of social behavior was found by Ruf and Fiedler (2000) in the caterpillars of the insect *Eriogaster lanestris* (Lepidoptera, Lasiocampidae). The same behavior was found in the tents of the pine processionary caterpillar *T. pityocampa* (Breuer and Devkota 1990). An increase in ambient temperature, such as the one predicted by global warming scenarios is going to have unpredictable consequences to the survival of the insect and its natural enemies.

The increase of the earth temperature has been associated with the loss of water resources. Some authorities believe that the creation of unpredictable aquatic habitats will promote population outbreaks and medical entomologists believe that the insect-borne disorders will become widespread through the migratory behavior of aquatic and semiaquatic insects (Epstein 2000). Since most aquatic insects are occupants of habitats occurring in the lowlands it is evident that the deterioration of these water bodies will cause extinction of insects incapable to migrate. Already Foster (1991) has reported the decline of *Emus hirtus* (Coleoptera, Staphylinidae) as a result of this inability. The decline is so subtle as to escape from inclusion in the British Red Data Book. As an example of the reshuffling of species as a result of the migratory behavior and the extinction of several native insects Petrakis (1991) reported several heteropteran insects of eremic origin in the coastal areas of Schinias, Marathon, Greece.

15.4 The prediction of certain insect extinctions

The prevailing opinion among scientists is that the extinctions are selective and affect only the most vulnerable species. For this reason with the appropriate design of reserves we can save the most endangered species by eliminating the factors that affect adversely these biological species (Lawton and May 1995; McKinney 1995; 1997; Purvis et al. 2000). Because the current insect extinction is much more intense –between 10^3 and 10^4 times than previous events (Purvis et al. 2000)– we can detect and measure the biological characters of the syndrome that make species susceptible to extinction.

15.4.1 The Darwin-Lyell extinction model

According to this model the [past] extinctions of single species and the phylogenetic clades where these species belong is a very slow process (Darwin 1859). In general, this follows the general lines that species as important biological units (Claridge et al. 1997) are subjected to a series of evolutionary events starting from origination and ending to extinction (Purvis et al. 2000). This scheme seems to have an impressive universality and has found support by many authors such as Gould et al. (1977) although some examples of sudden extinctions –or rapidly eliminated clades from the fossil record– apparently exist (Raup 1996).

Another important ecological process that Darwin accepted as a causal factor of extinction besides climatic stress is interspecific competition. This type of competition caused the selective extinction observed in a very poor fossil record (Raup 1996). In this record insects are, as a rule, very rare but they are categorized as not belonging to the extinction prone organisms for a variety of reasons. In this respect Whalley (1987) stated that the great mass extinction in the K-T boundary left unaffected the fossil record of insects (point 6 in Fig. 1) because these organisms recover fast and re-invaded the more affected areas. The tropics in this proposal were re-invaded from the northern or the Southern Hemisphere. The reasons for this are the high reproductive rate, the small size and the migratory ability of many insect species. Many current forms of insects are listed from the Cretaceous period and they continue to populate the layers in the Eocene and Maastrichtian while many forms exist today and they are abundant and effective competitors. Of course this cannot preclude that they will not succumb to the current extinction.

On the other hand many ecological processes in the ecosystems today promote intense competition among species such as the biological invasions of natural communities that act in many ways against native faunas and plant communities (Drake and Williamson 1986; Petrakis and Legakis 2005).

15.4.2 The Raup extinction model and the kill curve

Raup (1991, 1996) suggested an extinction model that overcomes the problems of the Darwin-Lyell, namely the passive replacement of faunas after extinctions and the different times required for a taxon to disappear from the geological layers. Passive replacement is used in the sense that the replacing fauna took advantage of the lack of the replaced one and did not compete directly with it. A direct competition would mean an active replacement.

The formulation of the extinction model involved the derivation of the "Kill Curve" (Fig. 4), which was validated against real world extinction records. The inventor of the model hopes to use it at shorter time periods and for the differentiation of background species loss from mass extinctions. As inferred from the frequency distribution of the extinction intensities it is clear that this pattern could not be explained by chance alone. In effect the extinctions were not independent events and the [pulses of extinctions of genera must be connected in some way ... because of common factors, such as ecological interdependence or shared physical stress]. The model in Fig. 4 is purely empirical and in fact describes in an easy way the sigmoidal curve of the Phanerozoic extinctions. The equation fitted by Raup (1996) is:

$$SE := \frac{\ln(WT)^a}{\left(e^b + \ln(WT)^a\right)} \qquad (15.1)$$

where SE stands for species extinction and WT is the variable describing the waiting time for these extinctions to happen. The quantities a and b were estimated by regression and were found to be 5 and 10.5. Raup validated the model with real and simulated data and found it to work at time periods not used in the formulation phase.

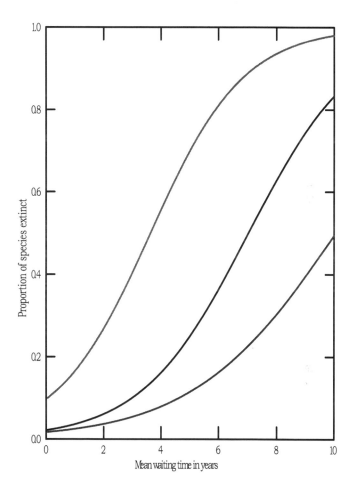

Fig. 4 The purely empirical quantal response Kill Curve of the Raup extinction model, together with the lower and upper 95% confidence curves. The curve is similar to the one presented in Raup (1996) though the character set was very restricted. The abscissa is log of years and the ordina is the ratio of extinct taxa

15.4.3 Causes of current insect extinction

Many authors have emphasized that our knowledge on the composition of insect faunas is fragmentary and incomplete. As a result the documentation of insect extinctions is very difficult unless we are willing [to spend more time and money] in order to study and document them (Dunn 2005). Samways (2006) in supporting the use of strategic indicators of species at risk says that the South African odonate insects *Pseudagrion citricola* and *Metacnemis angusta* changed conservation status when the methods of fieldwork and conservation practices were altered. More precisely, the first species was removed from the Red List when the new methods indicated that the population density estimation is greater than the one that dictated its conservation status. The second insect recovered and attained high densities when the invasive plants were removed from its biotope.

15.4.4 Climate change

The findings of some authors investigating the biological effects of climatic fluctuations in the glacial-interglacial cycles of the Pleistocene indicate that insects were largely unaffected. Coope and Wilkins (1994) state that if we restrict our interest to large invertebrates then we should predict that the extinctions of insects are widespread. However, the fossil record exhibits an impressive constancy posing a puzzle that is expected to affect our interpretations of past and future extinctions. Current extinction is not an exception from these predictions, especially in the lack of adequate knowledge on current insect faunas (Dunn 2005).

By using a species-area approach and the climate-envelope model with three climate projections:
I. Minimum scenarios that correspond to 0.8-1.7°C rise and 500 p.p.m.v. CO2.
II. Mid-range scenarios, which include 1.8-2.0°C and 500-550 p.p.m.v. CO_2
III. Maximum scenarios in which the temperature rise is >2.0°C while the concentration of CO_2 is elevated by an amount >550 p.p.m.v.

Thomas et al. (2004) predicted the percentage extinction of species in a set of regions that cover sixteen types of biomes in the Southern Hemisphere. I herein present only the data for butterflies, other invertebrates and the data pooled for all species, which were presented in a tabular format. The species that were used are coming from endemic species that have known geographical distributions. The species-area approaches were three and they corresponded to the following equation:

$$E_1 = 1 - (\sum A_{new} / \sum A_{original})^2 \qquad (15.2)$$

where E_1 is the proportion of species in a region that are destined to become extinct, $A_{original}$ is the area occupied by each species at present and A_{new} is the area predicted to be occupied at each one out of the three climatic scenarios. The summation is carried out over all endemic species of a region. This method actually estimates the *"overall changes in the areas of distribution"*.

The following equation estimates the *"average change in the areas of distribution"* of the species:

$$E_2 = 1 - \{(1/n) \cdot [\sum (A_{new} / A_{original})]\}^2 \qquad (15.3)$$

where n is the total number of species of a region and $A_{new} / A_{original}$ is the proportional change in the distribution of each species.

Then estimating the regional extinction risk is made by averaging it for all species in the region. For each species the extinction risk is given by the following equation

$$E_3 = (1/n) \cdot \sum [1 - (A_{new} / A_{original})]^2 \qquad (15.4)$$

in this equation if $A_{new} > A_{original}$ it was entered the analysis as $A_{new} = A_{original}$.

The fourth approach of Thomas et al. (2004) was not a species-area variant but it was received from Red Data Book(s). The criteria for such an approach were the ones used in RDB's. The authors used the logit-transformed three-way analysis of variance to estimate the extinction risk for each species. The criteria in RDB's expressed as a threat category, were used in the following Table 2.

Table 2. Criteria used in red data books

Threat category	Criteria	Extinction risk %
Not threatened	The decline in area in 50 or 100 years is small	0
Extinct	The projected area is zero	100
Critically endangered	The projected area in 50 years is < 10 Km^2 or the area decline is >80%	75
Endangered	The projected area in 50 years is 10-500 Km^2 or the area decline is 50-80%	35
Vulnerable	The projected area in 100 years is 500-2,000 Km^2 or the area decline is >50%	15

It is noticed that migrant taxa can decrease the risk of extinction most probably because they can select the most suitable habitat or can avoid several natural enemies or even can be an efficient competitor of introduced or invasive species. Also it can be seen that the butterflies can resist extinction more efficiently among other invertebrates and among all animals. This pattern is consistent among the methods and approaches of the estimation, among climate scenarios and among dispersing or non-dispersing taxa.

Even for the well-studied taxa like birds and mammals the estimation of the extinction risk is very uncertain (Heywood et al. 1994). The uncertainty stems from many sources the most important being [1] the exact meaning of the [simple] word extinction and [2] the doubtful inclusion of the fragmented habitats in the species-area relations used to estimate the extinction risk of species (Shaw 2005).

To illustrate the difficulty in the estimation of extinction risk and rates are presented the following examples, derived from the findings of current research. Traditionally the Mediterranean vegetation type is defined as the plant type that covers all areas with long summer drought and mild winters. Recently, Mitrakos (1980) presented convincing arguments that the Mediterranean climate is the one that is characterised by a long summer drought that poses a drought stress and a winter cold season, that poses a cold stress. Mitrakos devised two scales to measure the drought (D) and cold (C) stress that were independent (perpendicular) each other and are given by the equation below:

$$D = 2 \cdot (50-p) \quad \text{and} \quad C = 8 \cdot (10-t) \tag{15.5}$$

where p is the average month precipitation in mm, t is the average minimum month air temperature in centigrades. Calculating D and C for each month the monthly drought stress (MDS) and the monthly cold stress (MCS) are obtained. The summer drought stress (SDS) is obtained by adding the MDS of summer months (June, July and August) while by adding the MCS of winter months (December, January and February) the winter cold stress (WCS) is estimated. In southern regions (e.g. Chania, Crete, Greece) the stress imposed by summer drought is higher than the WCS while in colder regions (e.g. Florina, Greece) the reverse is true. In Fig. 6 on the $SDS - WCS$ plane are arranged several sites in Greece. Five groups can be readily recognised that are reflected to each one of the four Mediterranean tree-shrub species.

In a scenario involving temperature rise the swarm of points (=sites) is expected to be contracted to the left segment of the x – axis while on the y – axis it is expected to expand to the top well above the value 300 (Fig. 5).

In this climate scenario the arrangement of points ceases to be reflected to the existing floristic taxa. Trees and shrubs that behave well in the climate-envelope model locally will become extinct together with their specialist insect fauna –in both, taxonomical and ecological senses. Another extinction source is that the new climate will not have a common part with the climate envelope of several species and many insects among them. This last set is predicted to become extinct according to the algorithm of Thomas et al. (2004). The first source of extinction cannot be predicted by this and other species-area based models while it is usually

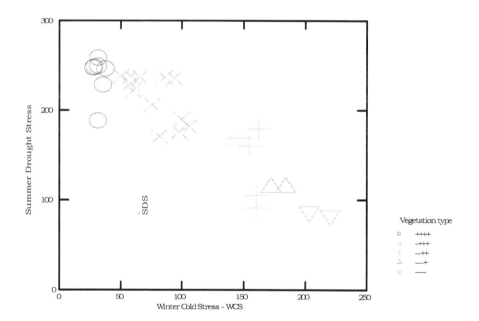

Fig. 5. Scatter-gram of meteorological stations from Greece. The plants that were used as indices are *Ceratonia siliqua, Pistacia lentiscus, Olea europaea, Quercus coccifera*. The five groups recognized in the scattergram are formed from these species in quadruples of signs according to their presences (+) or absences (-) in the sequence they are given here (after Mitrakos 1980)

called as co-extinction or habitat destruction / modification. Nevertheless it is a source of uncertainty to the anticipation of extinction rates.

Another problem is posed by many insect groups that exhibit a pattern departing even from the exceptional constancy of insect faunas reported by Coope and Wilkins (1994). When many models predict augmented insect extinctions Ribera and Vogler (2004) state that in the Mediterranean basin many Iberian endemic diving beetles (Coleoptera, Dytiscidae) originated in Pleistocene refugia. These refugia cradled the most important speciation of this insect group while another lower speciation event occurred in Late Miocene-Early Pliocene. These authors based their findings on the phylogenetic trees stemming from molecular sequence data of the genes 16S rDNA and cytochrome oxidase unit I (COI). On the basis of the finding that allopatric sister species are significantly younger than sympatric clades and that interglacial range expansion has not occurred in this insect group –i.e. diving beetles– the authors inferred that the allopatric model of speciation prevails. More precisely when species were categorised as 'allopatric' or 'sympatric' the differences in divergence time were significant in an one way ANOVA of both ultrametric trees under General-Time-Reversible (GTR) model –implemented in the program PAUP (Swofford 2002)– and the Non-Parametric-Rate-Smoothing (NPRS) model –implemented in TREEDIT computer program (Sanderson 1997). This contradicts the widespread opinion that Pleistocene is an epoch of very little divergence at species level. It was found that Iberian endemics speciated at this time. Probably this is in agreement with the findings in other insect species groups of the Northern Hemisphere and the authors give some similar works on the genus *Nebria* (Coleoptera, Carabidae) and Cicindellidae (Coleoptera) while they quote and one work on the post-Messinian (Late Miocene) speciation of brown frogs (Amphibia, Ranidae) (Coope and Wilkins 1994).

In addition, in the aforementioned work it is evident that it is very difficult to draw any sweeping generalization to infer speciation and extinction rates from widespread taxa. Actually the universality of this speciation paradigm in other Mediterranean areas rich in endemics remains to be discovered by additional work. For instance, in the analysis of Cretan endemics Legakis and Kypriotakis (1994) found that plant endemics form two distinct groups on the basis of the altitude of their distributions (high mountains –lowlands). Animal endemics in this study clustered in three groups with strictly geographical distribution in the western-central and eastern parts of the island of Crete. The authors state that this reflects the different strategies adopted by the endemic groups. Most importantly the past split of the island into three islands was differently perceived by taxa and in general animals do not follow plants in their distribution. It must be noted here that the endemic animals included also insectivorous taxa like the subspecies of gekoes *Cryptodactylus kotschyi* and the lizard *Podarcis erhardii*. Since then many insect endemics were found but this general pattern does not change. This analysis makes a departure from the species-area approach and supports the argument that insects merit a special approach that takes into account the idiosyncrasies of the taxa.

15.4.5 Habitat fragmentation, destruction, modification

In addition to the previous example, the work of Greuter (1972, 1979, 1995)–a systematic botanist–on island taxa indicated that island taxa, and insects predominate in these assemblages, exhibit some particularities. First, the earlier the habitation of the island by humans, as it happens in the Aegean and the Mediterranean in general, the lower the extinction rates unlike the islands in the Pacific where the human occupation varies inversely with extinction rates (Pimm et al. 1995). The first author speaks about plants while Pimm et al. tell us the story of birds. Birds adapt rapidly to human occupation and sparrows, starlings and doves are good examples of such an anthropophily while plants –and insects as well – are adapted to human settlements only through pre-adaptive characteristics and evolution of new traits. However, the evolution in the Aegean islands is almost at a standstill (Greuter 1979) and in this sense the island floras are of relict type except from some 'cryptoendemics', which form very sparse populations unlikely to be exploited by insects. The transgression of sea level in the interglacials lowered the temperature on the mountains by 6-7 °C and sea surface by 4 °C, which indicates that certain plant taxa will become extinct together with their arthropod faunas. Contrary to the findings of Ribera and Vogler (2004) for diving beetles, islands have never been evolutionary theaters and no plant originated in there since the Messinian salinity crisis. This entire region, Cardaegean region according to Greuter (1979) remained insular throughout Pleistocene sea transgressions. All these facts assure that extinction rates cannot be extrapolated in a geographical or a taxonomical sense, while in every case there is a sort of fragmentation of the previous geographical range of insect taxa.

Unlike the biogeographical fragmentation there is currently an ecological fragmentation at a considerably restricted scale, which is expected to affect first the populations of many species since these are the Evolutionary Significant Units (=ESU) (Crandall et al. 2000) and more drastically the stenoendemics. However, in many cases there is microevolution in many taxa that can be observable only at the level of – e.g. biochemical– markers (Petrakis et al. 2003). This is the case of *Hypericum empetrifolium*, a widespread phryganic and road verges plant, which gave rise to the eastern Cretan endemic *H. amblycalyx*, a calcareous cliff inhabitant, from which cannot be easily differentiated on a morphological basis (Greuter 1972). However, Petrakis and Park (2007), employing biochemical markers, have found that the latter species contains only a subset of the terpenoid complement of the first and this is related to though does not stem from the biotope it occupies. Actually the evolution in the genus *Hypericum* proceeds by eliminating terpenoids from the essential oils. This was found in many *Hypericum* species that inhabit cliffs and can be interpreted as a result of the lack of natural enemies and competitors in these habitats. This rendered the terpenoid defenses useless. Later these defenses disappeared by means of natural selection. Again, this transition from the biogeographical scale to the ecological scale of a cliff cannot be caught by the species-area based models. I am not aware from similar works on insects but this type of uncertainty is very likely to exist in estimating the numbers of extant insect species or extinct species in past extinctions.

On an ecological scale Steffan-Dewenter and Tscharntke (2002) by reviewing the works on the fragmentation of calcareous grasslands, stated that while small-scale studies dominate the literature there is a need for more large-scale studies. In a study on the role of small fragments of habitats to conserve insect communities Tscharntke et al. (2002) found that the habitats of the same area may harbor different numbers of species depending on the prevailing fragmentation. They worked with butterfly species of four degrees of specialization

- monophagous on one plant species
- oligophagous on one plant genus
- oligophagous on one plant family
- polyphagous on several plant families

It was found that the butterfly species-habitat area relations varied according to the category of specialization of the butterfly while for oligophagous insects the species-area curve was not statistically significant. In the case of the rape pollen beetle *Meligethes aeneus* (Coleoptera: Nitidulidae) they found that the parasitism was higher at the habitat edges where exist many overwintering sites of parasitoids –the natural enemies of the pollen beetle– and for this the fragmentation of the habitat may result in the extinction of the beetle and all species that are controlled by natural enemies.

This imbalance of the ecosystem procedures of the control of one species imposed by fragmentation is one action that suppresses biodiversity. The responses of a component of the ecosystem to changes in another component can be rapid regardless to the scale of the fragmentation (Zschokke et al. 2000). Other ecosystem functions that fail in fragmented habitats are related to the viability of insect species the most important being the restricted search space for food items and the loss of reproductive output due to the inability to find a mate. In this last case it must be said that in insects the finding of a potential mate is a difficult and error prone process and for this many adaptive strategies have been evolved such as sexual pheromones, conventional encounter sites and mating swarms. Except of some cases of sexual pheromones in strong flyers almost none of the strategies can be accomplished in a fragmented habitat.

Increasing the study scale of a habitat type the biodiversity is also increased and surrogates cannot adequately represent the biodiversity of the habitat type. In an attempt to augment the scale of study together with the ability to cover, for example, many types of grasslands, which comprise an important, rich in insect species and widespread biome in Europe, entomologists established a European network for Hemiptera in order to study their biodiversity (Biedermann et al. 2006). The same happens in other insect groups since the entomological work is taxonomically divided. Hemiptera is a good source of indicators and surrogates of the biodiversity of a habitat type that is rapidly modified due to climate change, land use status and alteration of management practices. For ground beetles and rove beetles the concern is urgent since they are adversely affected by habitat fragmentation and many works appeared to cover their utility in carrying out ecosystem functions, such as the population control of other organisms. More importantly, they are good inhabitants and indicators of ecosystem health in city parks and gardens. Starting from this point North American entomologists formed a group of scientist communicating on many aspects of ground beetle ecology and behaviour. On the event of the formation of the Midwestern Carabidologists Working Group (=MCWG) the journal American Entomologist of the Entomological Society of America published a series of papers in the journal section "Instant Symposium" (Lundgren 2004) where it can be found that besides their sensitivity to fragmentation, ground beetles are important bioindicators of a range of ecosystem processes and perturbations. Almost exclusively they are predatory on other arthropods and invertebrates in general and for this economically valuable. This is reflected in the number of papers that are published annually and are devoted to Carabidae. According to the AGRICOLA database in the years 2001-2007 there was an average of 122 publications per year devoted to Carabidae while many other publications do not exclusively contain ground beetles.

When a habitat is modified or lost the affected insect communities are expected to change. Probably the best studied examples are coming again from the most widespread habitat of the North Hemisphere the grasslands which are exploited by humans in the form of pastures. In these pastures the status has changed in general following the changes in livestock numbers. Roslin (2001) presented the course of livestock numbers in Finland (Fig. 6). By using various measures to estimate farm density, connectivity and size in a set of grid squares in which Finland was divided, Roslin was able to make a large-scale extrapolation of the previous local tests he made at the scale of two landscapes. The insect group used in this study was the dung beetle genus *Aphodius* (Coleoptera, Scarabaeidae), which forms assemblages that depend on the status of the pasture.

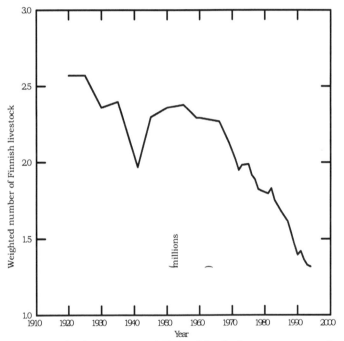

Fig. 6. The changes in the pastures of Finland in the last century are shown. The lowering of the curve in 1940 (World War II) is an intensification of the pattern of continuous decline (after Roslin 2001). Locally some pastures are expected to have extinct together with all specialist species of *Aphodius*

Roslin recognized four specialisation types in *Aphodius*; [1] 'forest specialists' that reproduce in woods although adults can be found in pastures; [2] 'generalists' that reproduce and feed both in pastures and woods; [3] 'intermediate species' that frequent both habitats; and [4] 'pasture specialists' that includes species almost exclusively found in this habitat. It was found that among the categories of specialization the most affected by the restriction of the habitat are 'pasture specialists', which declined and 'forest specialists' that increased. The other categories remained more or less unaffected although exhibited some variation. Within each category the member species were also variously affected. The species *A. sphacelatus* disappeared from the guild of intermediate species while the others remained more or less constant. The same disappearance occurred to *A. subterraneus*. Another finding of this research is that local biodiversities are always the result of a species reshuffling. This affected more than the longitudinal gradient of species richness and because of the reshuffling in order to compare local diversities they must always be conditioned to regional biodiversities.

The findings of dung beetle assemblages in Finnish pastures seem to be a realization of the species reshuffling prediction.

15.4.6 Species invasions / introductions

Many invasions have resulted in alteration of local faunas by eliminating some species and reshuffling the remaining ones. The same holds for introductions, which differ from invasions only in the agent of the transfer. Whereas in invasions the agent is ultimately the biological species itself armed with its adaptations, in introductions the agent is the man and always the introduced organism bears an economic benefit. This benefit ranges from simple ornamental or agricultural plants to classical biological control and apiculture, as is usually the case for insects. Many insects that produce honeydew, which is used by bees as food, have been introduced in several regions where they never existed before, and caused health problems to pine trees. In the case of insects the invasions are restricted to North America and Australia. Many insects were 'imported' in North America and caused extensive destructions of forests by severely modifying the habitat of many other species. Drake and Williamson (1986) reported that from the introduction of the gypsy moth *Porthetria (=Lymantria) dispar* almost a century later 600,000 hectares of forest have been defoliated while current predictions estimate the defoliation of forests to be 85 million hectares. Probably the defoliated plants have not become extinct but the habitat of many species is dramatically altered. At present we are not aware of extinct species because of this defoliation. It is very difficult to define exactly the range of affected taxa in the sense that Roslin (2002) done it.

At present only theoretical predictions and sparse experimental data are used to interpret the variability of insect responses (Tallamy 2004). Insects that face alien plants in their biotopes are predicted to be adversely affected since the invading plants are not random samples derived from the source biome. Always they are pest resistant plants with a special unpalatability to insects and in general conform to the industry demands such as the traits of introduced ornamental plants. For this very reason world floras are full of plants that were once introduced for ornamentation but escaped cultivation and became naturalized (e.g. Flora Europaea, Tutin et al., 1964-1980). All these invasive plants have become known only if they interfere with human activities and very rarely from botanic investigations and inventories. The impact on natural ecosystems is actually unknown and in this respect many insects have faced the alteration of their habitat with foreign plants that cannot be attacked / eaten. What are the reasons that many insect species have become extinct from this action remains unknown. Another reason for the success of introduced plants lies also in the fact that they are devoid of their natural enemy complex. The plant species *Phragmites australis* is an example and Tewksbury et al. (2002) reported that in Europe exists the Eurasian genotype that harbours more than 170 insect species. The same genotype introduced in North America harbours only 5 species of native insects. The same happens in the native (Australian) insect fauna of the tree *Eucalyptus stellulata*, which is eaten by 48 species of insects while in California it is not eaten by native insects (Morrow and La Marche 1978). Many other examples exist and the phenomenon is perceived by many culturists who are unable to distinguish these introductions from transgenic ornamental plants that pose an additional risk to native insect species. A third reason for the success of introduced plants lies in the arguments related to the co-evolutionary history. According to this (Tallamy 2004) native insects eat plants that have a common evolutionary history with the plants existing to their environment. Because of the arms-race type in the evolution of counter defenses, co-evolved insects bear the adequate counter mechanisms and substances to detoxify the defenses and finally consume the plants with which they have co-evolved. However, co-evolution requires evolutionary time and this is not given to the introduced plant species and the native insects. As a result the plants are propagated easily since there is no control exerted from the components of the ecosystem. Simultaneously the deprived insects as a result of competition and population control processes are not able to reproduce and probably they locally become extinct.

It is very strange that introduced plants do not attract insects even from the pool of specialists on their congeners. Novotny et al. (2003) working on a native and an introduced insect of *Piper* found that the second plant species harbored no specialists nonetheless the plant was introduced a century and a half ago. Turning to the effect that the introduction may have on generalists Tallamy (2002) presents convincing evidence that there is no any clear effect on the performance on native and introduced species. For Talamy the question remains open while Novotny et al. (2003) found no significant difference. This differential effect of the native and introduced plant species on specialist and generalist insect herbivores supports the third argument of the co-evolution of insects and plants. Generalists are designed to evolve a mechanism and the associated substances (e.g. mixed function oxidases) to deactivate the defence compounds of plants and for this they are able to exploit native and some introduced plants. Hence the confusing results found in the works quoted by Tallamy (2002).

On the other hand specialist insects had had not enough time to evolve such mechanisms and/or compounds so they do not belong to the insects harboured by the introduced plant. An outcome of this is the difficulty in finding specialist biocontrol agents for weeds that threaten local cultures and regional biodiversities. The introduced plants can be faced by another introduction from the source region. This was the case of introduced African dung beetles to contribute to the decomposition of non-marsupial dung in Australia, which the native insects could not decompose (quoted by Morris et al. 1991). We can easily anticipate that local insects will be at a shortage of food in the replacement of marsupials with cows. Again we are not aware whether some of the native insects became extinct. Certainly the introduced dung beetles thrived and cause a major reshuffling –and species extinctions?- of local insect communities.

15.4.7 Co-extinctions

Co-extinctions are more widespread than it was previously thought. In all paragraphs of this chapter it is explicitly stated that any action on the species of a local community affects many species. The number of affected species is strongly dependent on the connectivity of the community and the degree of its packing (Koh et al. 2004) while in some cases extinction is inevitable as is the case of the specialized parasites of an extinct host. The term co-extinction was initially coined to describe such systems (Stork and Lyal 1993). To estimate the number of coextinctions of species (=affiliates) that are so specialize as to exploit only one host species and become extinct because their host became extinct. Koh et al. (2004) developed a probabilistic model where they estimated the number of coextinctions from *'affiliate matrices'*. These affiliate matrices are simple matrices containing only 1's and 0's while the rows correspond to hosts and the columns to affiliate species. For systems where the host is exclusive for each parasite species the estimated function of extinction is linear while for systems where an affiliate species exploited more than one host the function was curvilinear.

Because, this model works only with full affiliate matrices, which means that the distribution of hosts must be known for each affiliate species, it is not applicable in many cases because of the lack of this knowledge.

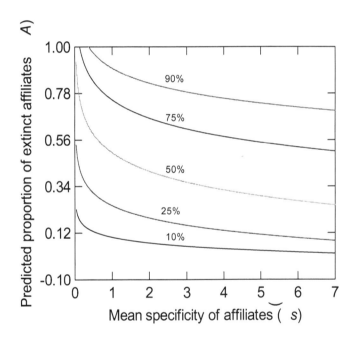

Fig. 7. The nomographic model of Koh et al. (2004) expressed as the empirical function $A=(0.35 \cdot E - 0.43) \cdot E \cdot \ln(s) + E$ where A is the estimated probability of extinction, s is the mean specificity of the affiliate species expressed as the number of exploited hosts and E is the probability of extinction of the host, which takes the values 10%, 25%, 50%, 75%, 90% to produce five lines of the same parametric family. Values of specificity above 8 (hosts) are almost horizontal indicating that the predicted proportion of extinction of the affiliates is not influenced by the extinction of a host. The model was estimated by means of 20 host-parasite systems, which includes Agaonidae wasps – *Ficus*, butterflies – larval host plants, Lycaenidae butterflies – ants

Also the authors developed a nomographic model (Fig. 7) to express the probability of extinction of an affiliate from mean values of specificity. Koh et al. (2004) state that *'mean host specificities'* are more easily estimated than *'complete host specificities'* and for this the nomographic model is more easily applicable.

It has been stated that organisms with complex life histories are expected to have higher extinction risks –because of coextinctions– than what is expected for species with simple ones. This is expectable since any obstacle to a certain step of the life history of the affiliate species would mean failure in reproduction. In a complex life history such steps are more likely to be found since they usually involve different hosts. In this way the quantity E increases since $E = \sum E_i$, where E_i is the extinction probability of the i-th host species. In general the extinctions of hosts are independent events. It is evident from the nomographic model of Koh et al. (2004) (Fig. 7) that the higher the E the higher the extinction probability of affiliates.

Also, it has been reported that a local extinction (Shaw 2005) of a host population causes the extinction of local affiliate population. Koh et al. (2004) quote the studied example of butterflies and their larval host plants and the state that this apart from the implications it may have to the species it also influences local conservation measures and management practices.

Many species extinctions, which have been reported under the heading habitat modification or human induced extinctions, are actually hidden coextinctions. Dunn (2005) reports some very interesting thoughts on this by quoting the example of a complex system involving the butterfly *Maculinea arion* that fed by workers and feed on the larvae of the ant *Myrmica subuleti*. Independently of this case in an attempt to control the population of the introduced rabbits (*Oryctolagus cuniculus*) conservationists in United Kingdom used strains of the virus genus *Myxoma*. The rabbits were actually controlled but the open habitats which were used –and created– by the rabbit for grazing also decreased in abundance. In effect the biotopes for the ant species were reduced and the butterfly *M. arion* disappeared, probably because it became locally extinct.

15.4.8 Hybridization and introgression

Hybridization and introgression are among the most difficult genetic phenomena that cause a mixing of the gene pools of distinct taxa (Rhymer and Simberloff 1996). Introgression is the gradual penetration of the gene pool of one species by another to form a series of hybrids (*hybrid swarm*) that in several cases goes unobserved (Petrakis et al. 2000). Hybridization is referred to individual organisms and is usually more easily detectable than introgression, which is the population outcome of hybridization. Very rarely introgression can be seen in morphological data especially if extensive backcrossing has occurred.

Although this kind of genetic mixing –i.e. introgression- has been studied extensively in fish populations it is suspected that it probably occurs extensively among plant species. Petrakis et al. (2000) have found that introgrssion occurs extensively between *Pinus halepensis* and *P. brutia* and the hybrid phenotypes. Because of this it is elicited a significant antifeeding behaviour to major pine defoliating insects, such as the larval stages of *Thaumetopoea pityocampa* and some bark beetles. In this way introgression is an important evolutionary process that generates new phenotypes within new scenery where defoliators suppress the intact pine populations whereas the produced phenotypes escape herbivory. Probably many insects cannot follow this arms race and do not adapt to the new phenotypes. Populations of this insect are destined to become extinct if the introgression proceeds to the entire geographical range of distribution. In this respect island populations and endemics are more vulnerable to extinctions since populations of a species are distributed in a restricted geographical range.

These two processes in natural situations act as a source of background extinctions. However, in many cases such as those studied by Petrakis et al. (2000) the substrate for introgression is created by man induced introductions.

Many conservationists have hypothesized that small range populations such as the populations on islands are more susceptible to extinction due to inbreeding. Importantly Frankham (1998, 2001, Frankham and Ralls, 1998) found that [many island populations showed levels of inbreeding correlated with elevated extinction rates]. For such populations introgressions may be an escape from inbreeding. Indeed, Rhymer and Simberloff (1996) state that small isolated populations are more likely to hybridize with juxtaposing previously isolated taxa. Such juxtapositions are very probable with human introductions of anthropogenic taxa, the increased mobility of humans either individuals or populations and the habitat modifications.

Hybridization between invasive and native species is considered as a factor of extinction. Usually native species do not function as evolutionary and ecological entities after the hybridization with an alien (=introduced or invasive) species. Such hybridizations have not been reported from insect invasions but several treatises of the theme discuss several instances in other terrestrial species. Cox (2004) states that in plants such hybridizations combined with polyploidy gave rise to a number of new species, as is the case of the introduction of European species of *Tragopogon* into North America. The same happened in freshwater fishes, crustaceans and molluscs.

In general all the causes of extinction discussed under separate headings –habitat destruction / modification, hybridization / introgression, species invasions / introductions– as a rule are connected and act synergistically to cause species or ESU extinctions. Any attempt to separate them is solely for the facilitation of discussions.

15.5 Conservation of insect species and their biodiversities

The current biodiversity crisis is unprecedented in the biotic history of earth both in time and numbers of taxa destined to become, extinct. The reaction of many scientists and politicians is to try to stop or reverse this process or mitigate the effects that are expected to have on human societies. All people agree that for certain areas or taxa a priority is required since not all species can be the targets of conservation. This necessity triggered many articles and attempts to device selection methods for species, faunas, and areas of high biodiversity.

15.5.1 Estimating conservation priority

A problem with insects is that in past extinctions they are among the least suffering organisms. McKinney (1999) states that two biases are usually responsible for the underestimation of insect extinctions. Insects are an understudied group of organisms with some abundant taxa but many sparse populations with densities on or below the threshold line that make them perceptible to scientists. [1] There is the tendency to record common-species and for this rare species with sparse populations go unrecorded; this bias is named '*common-species bias*'. [2] The effort devoted to the recording of sparse and rare species is minimal and for this the extinction or the threat status of these species is dubious; this type of bias is called '*evaluate neglect*'. McKinney presents a good example of these biases concerning butterflies. These insects are among the pronounced insect species and for this they are included in Red List(s). Even in this case the inclusion of 159 threatened species among the 15,000 described species worldwide is particularly small (1.06%) far less than the amount of mammals (24%) and birds (11%).

One way to improve the estimation of biodiversity is to improve the measurement methods. McKinney developed the set of equations – described below– a diagram of which is shown in Fig. 9. The equation attempts to model the fact that the existing methods are not devised for all species and the widely used proxy group for biodiversity estimation, i.e. mammals, may be not adequate to account for groups such as insects (Fig. 8).

In the model $p \cdot k$ denote the probability of extinction of the rarest species; the extinction of the next rarest species is $p \cdot k^2$ and the extinction probability of the third rarest species is $p \cdot k^3$. In these quantities p is the probability of extinction characteristic to the group of species and k is the decrease in extinction of the next species. If $k<1$ the probability of extinction is modelled to increase with increasing rarity. Under the assumption that species follow the log series pattern without loss of generality the sum of all species expected to become extinct is $p/(1-k)$ while their fraction in an assemblage of n species is $p/[n(1-k)]$. If u represents the unrecorded species then the fraction of recorded species is $(n-u)/n$. The extinction probability of the u-th species is $p \cdot k^u$ and the fraction of extinctions for a recorded species are $p \cdot k^u/(n-u) \cdot (1-k)$. If the y-axis shows these values and the x-axis holds the values of the fraction of recorded species then Fig. 8 depicts the curves for three groups of organisms, i.e. mammals, plants and insects. Table 3 contains the parameter values for each group.

Table 3. Parameter values for each group of organisms			
Goup	$p_{lower} - p_{upper}$	k	N (x1000)
mammals	0.1 – 0.9	0.999	5
plants	0.1 – 0.9	0.9999	250
Insects	0.1 – 0.9	0.999995	8,000

The conservation of insects started only recently and the first attempts tried to prove that insects deserve conservation (all the papers in Collins and Thomas, 1991; Kim 1993a, b). Other studies directed towards the device of indices that estimate biodiversity taking into account all the biologically important aspects. In an important paper Vane-Wright et al. (1991) found that the existing biodiversity indices that base their importance in including richness and abundance. They devised a novel index based on the information content of the cladistic classification of the target group. This index includes all the information content of the hierarchical classification. As stated by Vane-Wright et al. (1991) the index is also dependent on the specific topology, hopefully the phylogeny if cladistics was used for the construction of the tree. We reproduce their example in Fig. 9 by assigning the standardized weights to the species of the phylogeny represented by the dendrogram.

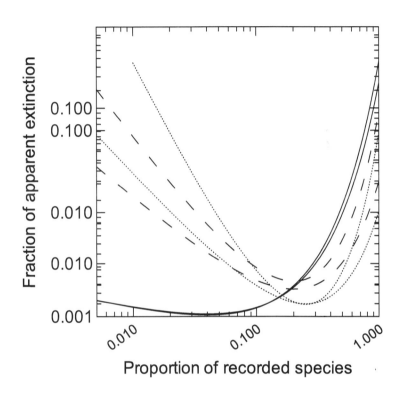

Fig. 8. Curves for the relationship between the proportions of apparently extinct species with the proportion of recorded species on the basis of the equation derived by McKinney (1999). Six curves are drawn, two for each taxonomic group, i.e. mammals, plants and insects with the probability of extinction for each group as $p=0.1$ and $p=0.9$. Continuous lines are for mammals, dashed lines are for plants and dotted lines are for insects. Further explanations are given in the text

In three regions there were found five species. For each species is assigned a weight on the basis of the phylogeny shown in Fig. 9. The phylogenetic importance of each species is expressed as the number of nodes met on the tree starting from the species and ending on the root of the classification. Then each weight is expressed as the standardized proportional score calculated by dividing each value with the sum of weights. A second standardization follows by dividing each species weight with the largest weight. Schematically the double standardization and the production of the weight (W column) is (A, B, C, D, E) → (4, 4, 3, 2, 1) → (3.5=14/4, 3.5, 4.67, 7, 14) → (1, 1, 1.33, 2, 4) = W.

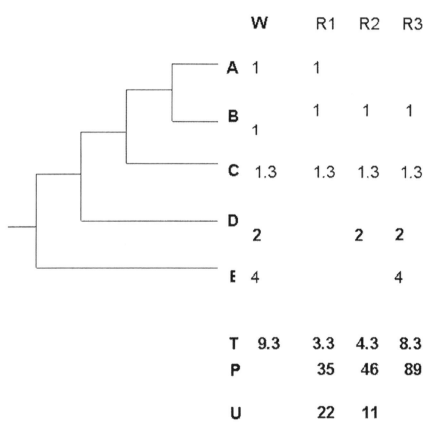

Fig. 9. The design of reserves takes into account the phylogeny of the species expressed as a standardized species weight (W), which modifies the presence of each species in each of the three regions (R). T is the total of weights, P gives the percentage diversity scores for each of the three regions and U gives the percentage diversity scores on the basis of the complementary taxa to region R3 (Vane-Wright et al. 1991) (see text for explanations)

This weight is put in the place of 1 (=presence) in each of the three regions R_i. In this way are constructed the matrices W(5 x 1) and R(5 x 3) as shown below:

Species(1..5) Regions(1..3)

$$W := \begin{pmatrix} 1 \\ 1 \\ 1.3 \\ 2 \\ 4 \end{pmatrix} \qquad R := \begin{pmatrix} 1 & 0 & 0 \\ 1 & 1 & 1 \\ 1.3 & 1.3 & 1.3 \\ 0 & 2 & 2 \\ 0 & 0 & 4 \end{pmatrix}$$

The quantities T and P in Fig. 9 have been computed below, together with the numeric example. The values in the matrix P (3 x 1) are the species presences weighted by species weights. The third region with the highest score –i.e. 89– is judged as the first to be considered in conservation programs. The second in consideration is R_2 having a score c.46 while the first region is R_1 as having the lowest score that is 35. The ranks of regions R1 and R2 can change if the scores are computed by taking into account only the species unique to each region –U scores. The region R_1 (score 22 in Fig. 9) has one unique species not occurring in other regions while region R_2 (score 11) shares the third species with region R_3. For that reason R_1 is preferred for conservation.

$$T := \sum W$$

$$P := 100 \begin{pmatrix} \dfrac{\sum R^{\langle 1 \rangle}}{T} \\[6pt] \dfrac{\sum R^{\langle 2 \rangle}}{T} \\[6pt] \dfrac{\sum R^{\langle 3 \rangle}}{T} \end{pmatrix} * \qquad P = \begin{pmatrix} 35 \\ 46 \\ 89 \end{pmatrix}$$

This algorithm has been implemented in the computer package WORLDMAP of Williams (1995, 2000 for a more recent version of the software [a demo is downloadable from the same site]). The program is modified as to include the algorithm depicted in Fig. 9 in a '*priority area analysis*'. The *Bombus sibiricus*-group with a known phylogenetic diagram is included as a real world example for this program (Vane-Wright et al. 1991); it includes 43 species in a set of 120 out of 250 grid squares in which the author divided the earth. Equador with 10 *B. sibiricus* species contributed by less than 15% to the total percentage diversity. The area Gansu, China with 9 species contributed by 23%. He found that the area Big Horn, USA with 4 species, has higher complement from Gansu and contributes more than 15% to the total percentage biodiversity. On the same basis Equador is complementary to both Gansu and Big Horn and contributes by 14% to the total biodiversity. So, these three squares –i.e. Gansu, Big Horn, Equador [in this order]– contribute by 52% to the total (world) biodiversity of the *B. sibiricus*-group. Should only two areas must be selected for conservation Gansu and Big Horn must be preferred on the basis of the complementarity principle.

As a criticism for the use of WORLDMAP, Harvey (1995) describing the 1995 annual meeting of the Zoological Society of South Africa, discusses the applicability of WORLDMAP program in the design of protected areas for non-passerine birds in tropical Africa. He concluded that the program produces measures of taxonomic dispersion and not measures of genetic diversity. Moreover, according to the same author it is not "clear why genetic diversity (measured as branch length in a phylogenetic tree) is equally weighted with richness". Continuing, Harvey reported that in South Africa the protected areas are so designed as to save many endangered endemic species and not the maximum number of species (maximum richness) or the maximum number of endangered species, which are abundant or well represented in other parts of Africa (Harvey 1995).

Many other scientists criticized the approach of Vane-Wright et al. (1991) the most important being that of Crozier (1992). The criticism was expressed on the basis that the Vane-Wright et al. algorithm is specific to the topology of the inferred phylogeny. This topology can be either untrue one or even may be unavailable, which is more frequent. In addition it does not take into account other criteria such as the non-biological reasons to protect a place, a habitat or an organism. Crozier (1992) supports in replacement of the method of Vane-Wright et al. another algorithm based on the genetic distance as a priority criterion. Later (Crozier 1997) he stated that the number of genes is a safe criterion to assess the conservation priority of an entity (habitat or taxon). The method suggested by Crozier avoids many obstacles of the previous algorithm such as the number of nodes and the branch lengths in the phylogram and exploits the information content of the genomes of candidate for conservation taxa. According to these arguments the vertebrates having c. 80,000 genes deserve conservation better than insects incorporating c. 20,000 genes. Various workers selected different genetic material to analyze apart from whole genes. The advent of PCR, which is capable to augment very tiny DNA samples, made possible the investigation and enumeration of previously unknown bacterial biota by identifying bacterial species from complete genes. Another method through which the Guanine + Cytosine content of the sample is measured the number of species can be estimated. At what degree these results can be transferred to insects at present is still unknown.

Others recognize that the genetic diversity is an important part of biodiversity and believe that estimating the conservation value of a set of areas or a set of species is substantial, which the method of Vane-Wright et al. (1991) seeks in interspecifically estimated phylogenies. Moritz and Faith (1998) argue that estimating conservation value of species and areas from intraspecific or population phylogenies derived from molecular data is a much better identification of the suites of species that assure the maximization of 'feature diversity'. Posadas et al. (2001) accepted the measure of Vane-Wright et al. (1991) but they state that it has to be improved as to include complementarity and endemism. They suggested a set of indices that they succeeded in doing so at various degrees and presented an example from twelve South American sites inhabiting eight genera of insects (Coleoptera, Carabiidae and Curculionidae), 3 genera of spiders (Araneae, Gnaphosidae, Nemessidae, Anyphaeniae) and 2 plant genera (Plantae, Asteraceae). They calculated the indices (I, W) (=raw indices), (I_s, W_s) (=standardised indices) (I_e, W_e) (=endemicity indices), (I_{es}, W_{es}) (=endemicity standardised), *Richness* and *Endemicity* from the respective values of species in the known phylogeny of each genus. In Fig. 10 are presented some aspects of the computations on W and the calculation of the values for each area on the basis of the presence of each species. Details for the computations of the other indices can be found in the paper of Posadas et al. (2001). The most important finding is the change in rank order of areas according to various indices. The standardisations of the indices I, W, I_e, W_e dramatically changed the ranking of the set of twelve areas. This means that according to the index used for the ranking of the candidate conservation areas change priorities. Also, the fact that the phylogenies must be known is a disadvantage of the method, which greatly impedes its applicability. Fortunately, many people provided new ideas, methods and tools to overcome these problems (Harvey 1995; Williams 1995).

Faith et al. (2004) commenting on the approach of Posadas et al. (2001) state that this method is actually dangerous because it constructs a mixture of correlates with phylogenetic diversity such as complementarity and endemism. They concluded that although the approach of Posadas et al. (2001) is directed towards the inclusion of historical information in estimating conservation priorities it fails because of the poor knowledge of definitions of terms such as diversity, endemism, and complementarity. In fact, the employment of a cladogram or a phylogram does not guarantee that historical information is included in the conservation of biodiversity. Faith et al. (2004) are right when say that the use of intriguing terms with many interpretations and definitions may result in "an unworkable framework" of theoretical predictions.

Many species or habitats are protected not because of their high biodiversity or the low rank they attain in a conservation priority value but simply because they are or contain endangered species such as the rhinoceros or the birds of paradise. Many insect species that inhabit the conserved areas are also conserved as companion species. However, all these places include only a small fraction of the insect biodiversity in any sense. Moreover, it is unlikely that any insect species can bear the cultural load of a rhinoceros or the birds of paradise for human societies (Crozier 1997).

15.5.2 Conserving insects through habitat / plant conservation

Plants give biotopes their physical configuration and for this it is expected that they will be correlated with several aspects of insect species richness. It has been found in several studies that the architecture and the associate taxonomic richness of plants considered both individually and collectively as a vegetation cover (=plant community) is correlated with the number of harbored insect species (Lawton and Schroeder 1977; Sieman et al. 1998; Menendez et al. 2007). Recently Panzer and Schwartz (1998), in a very important paper, analyzed a set of nine plant measures in fifty prairie reserves in North America. They correlated these measures with aspects of insect richness. Three of these measures, namely plant community richness, plant species richness and plant genus richness, were significantly correlated with the richness of insects. Importantly, the plant family richness was never significantly correlated with insect species richness. This seems to be germane in some way to the past extinctions of insects, which are usually studied at the family level. Rare plant species were similarly insignificant to insect species richness, which is somewhat surprising because rare plant species assure the protection status and conditions of the reserve. They concluded that the conservative way to design protected areas in order to save many insects from extinction must be based on the detailed knowledge of the biology of the participating species, which is not possible or feasible in the real world. Plants are usually a good indicator taxon –proxy group– that assures the conservation of vertebrate species, and insects as well in the sense of a side effect. The inclusion of rare insect species did not coincide either with plant richness or common insect richness. This supports earlier suggestions against the use of the diversity "hot spot" approach in the selection of conservation areas. This approach might fail to capture many rare insect species. The area of the reserve was weakly correlated to the richness of rare butterflies ($r = 0.212$, $P < 0.01$), rare leafhoppers ($r = 0.423$, $P < 0.05$), and rare butterflies and leafhoppers pooled ($r = 0.518$, $P < 0.01$); when the area was combined with plant community richness the correlation coefficients slightly increased. In all other cases they were insignificant.

Petrakis (1991) found a weak though significant correlation among plant and insect diversity. The same author, studying the tritrophic complex involving plants, insects and birds found that the geographic origin of bird species was the reason for the selection of closed biotopes by birds that frequent mid-European ecosystems (Petrakis 1989). On the other hand Mediterranean birds prefer open biotopes. For that reason the vertical structure of the biotope –i.e. verticality– is very well correlated with the *'forest'* group of insects. For the Mediterranean group of birds the horizontal structure –i.e. horizontality– is the important biotope parameter that affects bird preference. In all cases of birds the succession stage of the vegetation was also important mainly because at that stage the vegetation harbors the highest possible insect diversity and abundances, which implies enough food for insectivorous birds. Indeed, in another study (Petrakis 1991) I found that in an East Mediterranean ecosystem the season with abundant resources for plants and in effect for insects –i.e. spring– causes the appearance of the correlation between plant and insect diversities. In advanced stages of the succession, plants harbour a predictable and important arthropod (and insect) fauna (as it is shown by the works of Ranius 2002 and Ranius et al. 2005 for *Osmoderma eremita*). This particular fauna is generally preferred by birds although some bird species found better hunting biotopes at the margins of forest openings where many insect species exploit the high plant diversity. This last group of birds does not contain a high proportion of insectivorous birds since the openings are dominated by annuals that early in the season produce many seeds, which are suitable food for granivorous birds.

The habitat type is expected to affect the diversity of captured insects. Hughes et al. (2000) examined three habitat types meadow, aspen and conifer in four geographic areas by means of Malaise traps. They found mainly Diptera and Hymenoptera, 8847 and 1822 individuals respectively while 2107 individuals belonged to various other orders. All Diptera and Hymenoptera species were identified to the family and morphospecies level. It was found that Diptera and Hymenoptera were significantly differentiated by habitat type. Importantly, when a cluster analysis was performed by means of additive similarity trees on the basis of Diptera captures of Malaise traps in the twelve sites (four areas) x (three habitat types) - the produced dendrogram was almost perfect and the three habitat types were clustered in the same group except for the aspen and conifer types, which were clustered together in a separate group. Because the authors examined many correlations, and found the significant correlation of Diptera with Hymenoptera with the number of sampled insect orders were it is assumed that among insect orders many surrogates can be found. Also, it was found that the feeding guild and the body size of each insect were important determinants of the habitat specialization.

Many predictions involving habitat specialisation process data are derived from adult insects captured in a set of traps. However, as Hughes et al. (2000) state, there are many problems with the measurement of habitat specialisation. The most important problem is that the habitat of adult and larval stages can be different. They quote the example of Syrphidae (Diptera), which are predacious at the larval stages but nectarivorous or pollen feeders as adults (Chambers and Adams 1986).

In several cases the conservation is taxonomically and geographically restricted for non-biological reasons as in the protected areas of western Crete. The report of Thanos et al. (2005) describes a set of seven plant (only) reserves out of fourteen important sites on the island of Crete where only endemics, rare and threatened plant species are protected. The list of protected plants is given in the abstract of Thanos et al. (2005). There is no biological reasoning to support all these selections. The pronounced administrative / economical criteria were [1] the small total protected area emanating from the word "micro-reserves" and dictated by the demand of pastures for sheep and goat; [2] the connectivity of the various small preserves emanating from the word "network"; [3] the dispersion of mountainous and coastal reserves so the tourist enterprises are homogenously benefited; [4] the establishment of monitoring devices and the required personnel is homogenously distributed so all local people are benefited from the reserves.

15.5.3 Conserving insect biodiversity in city parks and road verges

Among the human structures that fragment in a definite way the habitat of many species are the roads (Saarinen et al. 2005). Many animals attempting to cross this barrier are being killed by traffic. For the United States it is estimated that the daily rate of road-killed vertebrates is one million (Ries et al. 2001). No similar rates for invertebrates have been estimated but this source of mortality is of minor importance compared to other mortality sources. When Ries et al. (2001) followed 108 butterflies (individuals) in Iowa and reported that the material or the texture of the surface of the road had no effect on the behaviour of butterflies while the significance of an accidental road-kill was practically nil since only three of them had a collision with a vehicle. The overall mortality risk because of the road was very low (2.8%). The same result reached by Monguira and Thomas (1992) for butterfly and burnet populations in Europe.

In many cases the road verges are areas receiving high pesticide load from nearby fields. Also the traffic alone can impose poisonous contaminants such as heavy metals, to all organisms living in the verge near the road (Riemer and Whittaker 1989). As an additional habitat for native plants and insects the road verges can potentially serve important conservation purposes (Ries et al. 2001). It seems that conservation is more important for plants, especially native grasses and prairie species, while for weeds, there is a strong concentration to road verges. In Fig. 10 is shown the habitat preference of the two ecological types of twenty-five butterfly species used by Ries et al. (2001) i.e. the disturbance-tolerant species and the habitat-sensitive ones. Many important points useful for conservationists emerged from this study. The habitat-sensitive butterflies were much fewer than the disturbance-tolerant ones. Among the latter the weedy and prairie verges were not different. Expectedly, habitat-sensitive butterflies had significantly different abundances in all types of verges. In the broader region where the study site of Ries et al. belongs –i.e. Iowa– the tallgrass prairie is at severe decline and topically the habitat losses vary among 82 and 99%. The road verges must be properly managed in order to present plant species and endangered butterfly species additional habitats. Further details can be found in the paper of the team.

The city habitats either as suburban backyards or as parks in the city centre are also important habitats for a myriad of insects such as butterflies and dragonflies apart from other beneficial organisms such as ground and rove beetles and non-flying insectivorous mammals. In the park 'Antonis Tritsis' can be easily observed individuals of dragonfly species supervising their territories in the artificial biotope of a lake near Mt Poikilo, Attica,

Greece while pitfall trapping at Alsoupolis, Philothei and Chalandri, Athens, Greece gave several Carabidae and Staphylinidae (Coleoptera) species some recorded for the first time from Attica or Greece.

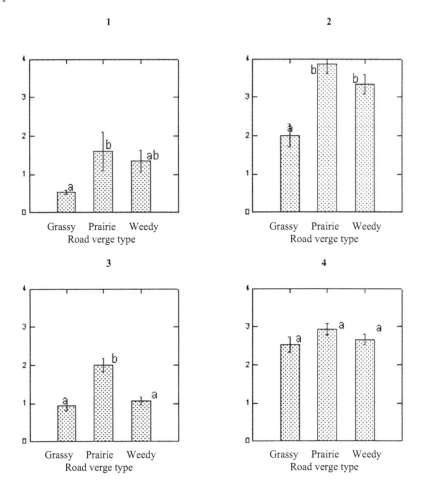

Fig. 10. Mean abundance and mean richness of butterflies in the three types of road verges that is Grassy, Prairie and Weedy. The digits above the diagrams denote 1, 2: mean abundance and 3, 4 mean richness of butterfly species. The error bars correspond to one SE. Different letters denote significant difference at probability levels <0.02, <0.01, <0.0001 and 0.31 (after Ries et al. 2001)

These numbers almost zeroed when spraying with insecticides was applied for the control of pine scale *Marchalina hellenica* (Hemiptera, Sternorrhyncha, Margarodidae). The same happened at various places in Philothei and Chalandri at the spots where pitfall traps were set (Mita et al.

2002; *personal observation* and unpublished data). Whether these treatments of city and suburban habitats with insecticides results in a loss of species remains to be discovered with additional studies since insecticide applications are usually done in countries with largely unexplored insect faunas. For the mentioned suburbs of Athens, Greece these extinctions have high possibilities. Many endemic species exist, while city parks and road verges are full of native plant species harboring an important entomofauna since no fertilizers are normally applied (Munguira and Thomas 1992)

Deichsel (2006) investigated the effect of the urbanization on the communities of ground and rove beetles (Coleoptera, Carabidae and Staphylinidae). Although the ecology of these beetles is not known at depth they are appreciated by experts as control agents of many parasitic invertebrates. Especially, in Central Europe they are important components of forest ecosystems and they are effective indicators of urbanization effects. Deichsel found that in Berlin, Germany rove beetles are important bioindicators of fragmentation, degradation and heating due to urbanization effects. These beetle assemblages predominantly comprise flightless insects. The prevailing explanation is that the fragmentation of the habitat associated with urbanization releases a multitude of adverse conditions to insect inhabitants, which react either by migration –the flying forms- or by local extinction. However, Deichsel (2006) did not find differences in richness across rural to urban gradients, although the specialist species –e.g. forest specialists- are absent from sites with apparent signs of urbanization. The salient result of this and other related studies is that urbanization promotes the epigeic insect species that are resistant to the disturbances associated with urbanization while specialist species are locally extinct. It is expected that insect extinctions of species with reduced geographical ranges will result in irreversible extinctions. It seems that this is a part of a more general rule, which states that the richness of various components of the ecosystems are associated in a more or less well defined correlation (Wolters et al. 2006).

In a detailed study of the carabid assemblages (Coleoptera. Carabidae) in brownfield habitats –post industrial derelict wasteland– Small et al. (2003) found that two scarce species of open habitats occur in these sites among the 63 species of the entire study. Moreover, 12-15% of the UK Carabidae species occur in brownfields. However, the investigators noticed that the conservation of these habitats and their richness is problematic because [1] of the rapidity with which the succession proceeds towards less rich in carabids land communities and [2] these sites were proposed by UK governmental authorities as candidate sites for residential redevelopment. The importance of the conservation of early succession habitats created in densely populated areas of Europe and their ability to harbour rare or threatened insect species is exploited in limestone quarries (Benes et al. 2003). This type of habitat can be used for reserve purposes for many xerophilous organisms. Indeed, many xerophilous butterflies inhabit quarries. In a suggestion of Benes et al. (2003) the creation of quarries is a method for creating suitable habitats for butterflies provided that there is enough heterogeneity of habitats and there are enough species of such insects in the vicinity. Apparently quarries are considered as escape sites for such butterflies and for some authors the quarries are rehabilitated not only for aesthetic purposes but also for maintaining local biodiversity and stopping insect species extinction because of habitat loss (Brofas 1992).

Again these habitats can be important escape sites in places where the agricultural land with the associated intensification and the exploitation of marginal land pressed many insect species to migrate or become extinct.

15.6 Acknowledgements

Professor Vassilios Roussis, Dr Lanna Cheng and Assistant Professor Constantinos Vagias are thanked for their helpful discussions on an early draft of this paper. Dr Ralph Lewin made useful corrections on an early draft of the manuscript. Many people helped with the literature, library and computer (database) work among them Mrs Emily Lahlou and Mr D.K. Roumeliotis are greatly thanked. Money for this work came from the Greek Secretariat of Research and Technology Contract no KOR-017.

References

Barber VA, Juday GP, Finney BP (2000) Reduced growth of Alaskan white spruce in the twentieth century from temperature–induced drought stress. Nature 405: 668
Beerling DJ, Woodward FI (1996) Palaeo-ecophysiological perspectives on plant responses to global changes. Trends in Ecology and Evolution 11:20-23
Benes J, Kepka P, Konvicka M (2003) Limestone quarries as refuges for European xerophilous butterflies Conservation Biology 17: 1058-1069
Benton MJ (1990) Vertebrate palaeontology. Boston: Unwin Hyman pp377
Biedermann R, Nigmann U, Achtziger R, Nickel H, Stewart AJA (2006) Effects of land use change on the biodiversity of Hemiptera in grasslands: a pan-European perspective (a summary). Hemiptera Ecology Network, 2006 Workshop, Laufen/Salzach, Germany

Boettner GH, Elkinton JS, Boettner CJ (2000) Effects of a Biological Control Introduction on Three Nontarget Native Species of Saturniid Moths. Conservation Biology 14: 1798-1806

Breuer M, Devkota B (1990) Studies on the importance of nest temperature of Thaumetopoea pityocampa (Den. & Schiff.) (Lep., Thaumetopoeidae). J.Appl.Entomol. 109,331-335

Brofas GL (1992) The effect of soil depth to the survival and growth of Alepo pine seedlings planted on quarry terraces in Drakia, Magnesia, Central Greece. Proceedings of the 7th Hellenic Edaphological Congress Larissa, Greece, pp 419-429

Cabeza M (2000) Habitat loss and connectivity of reserve networks in probability approaches to reserve design. Ecology Letters 6: 665–672

Chambers RJ, Adams THL (1986) Quantification of the impact of hoverflies on cereal aphids in winter-wheat: an analysis of field populations. Journal of Applied Ecology 23: 895-904

Claridge MF, Dawah HA, Wilson MR (1997) Practical approaches to species concepts for living organisms. In Claridge MF, Dawah HA, Wilson MR [editors] " Species: The Units of Biodiversity Chapman and Hall, London pp 1-15

Claridge MF, Gillham MC (1992) Variation in populations of leafhoppers and planthoppers (Auchenorrhyncha): Biotypes and biological species. In Sorensen JT, Foottit R (eds) Ordination in the Study of Morphology, Evolution and Systematics of Insects. Elsevier, Amsterdam, The Netherlands 241–260

Colinvaux PA (1973) Introduction to Ecology Wiley, New York pp 1-621

Collins NM, Thomas JA (eds) (1991) Insect Conservation. Academic Press, London

Coope GR, Wilkins AS (1994) The response of insect faunas to glacial-interglacial climatic fluctuations. Philosophical Transactions of the Royal Society of London B Biological Sciences 344: 19-26

Cox GW (2004) Alien Species and Evolution: The Evolutionar Ecology of Exotic Plants, Animals, Mivrobes, and Interactiong Native Species. Island Press, Washington D.C. pp 1-377

Crandall KA, Bininda-Emonds ORP, Mace GM, Wayne RK (2000) Considering evolutionary processes in conservation biology. Trends in Ecology & Evolution 15: 290-295

Crozier RH (1992) Genetic diversity and the agony of choice. Biological Conservation 61: 11-15

Crozier RH (1997) Preserving the information content of species: Genetic diversity, phylogeny and conservation worth. Annual Review of Ecology and Systematics 28: 243-268

Darwin C (1859) On the Origin of Species by Means of Natural Selection. John Murray, London

Deichsel R (2006) Species change in an urban setting-ground and rove beetles (Coleoptera Carabidae and Staphylinidae) in Berlin. Urban Ecosystems 9: 161-178

Dempster, J. P. (1991) Fragmentation, isolation and mobility of insect populations. In "The conservation of insects and their habitats" Academic press, San Diego pp.143-153

Diamond, JM, May RM (1981) Island biogeography and the design of natural reserves. In May RM (ed.) "Theoretical ecology - principles and applications" Blackwell Scientific Publications, Oxford pp 228-252

Dobson A, Lodge D, Alder J, Cumming GS, Keymer J, McGlade J, Mooney H, Rusak JA, Sala O, Wolters V, Wall D, Winfree R, Xenopoulos MA (2006) Habitat loss, trophic collapse, and the decline of ecosystem services. Ecology 87: 1915-1924

Drake JA, Williamson M (1986) Invasion of natural communities. Nature 319: 718-719

Dunn RR (2005) Modern insect extinctions, the neglected majority. Conservation Biology 19:1030-1036

Epstein PR (2000) Is global warming harmful to health? Scientific American 20 : 19-23

Erwin DH (1992) A preliminary classification of evolutionary radiations. Historical Biology 6: 133–147

Erwin DH (1998) The end and the beginning: recoveries from mass extinctions. Trends in Ecology & Evolution 13: 344-349

Erwin DH Valentine JW, Sepkoski JJ (1987) A comparative study of diversification events: the early Paleozoic versus the Mesozoic. Evolution 41: 1177- 1186

Faith DP, Reid CAM, Hunter J (2004) Integrating phylogenetic diversity, complementarity, and endemism for conservation assessment. Conservation Biology 18: 255-261

Foster DR, Oswald WW, Faison EK, Doughty ED, Hansen BCS (2006) A climatic driver for abrupt mid-holocene vegetation dynamics and the hemlock decline in new England. Ecology 87: 2959-2966

Foster GN (1991) Conserving insects of aquatic and wetland habitats, with special reference to beetles. In "Conservation of Insects" Collins NM, Thomas JA (eds) "Conservation of Insects". Academic Press, London, 238-260

Frankham R (1998) Inbreeding and extinction: island populations. Conservation Biology 12: 665-675

Frankham R (2001) Inbreeding and extinction in island populations: reply to Elgar and Clode. Conservation Biology 15: 287-289

Frankham R, Ralls K (1998) Inbreeding leads to extinction. Nature 392: 441-442

Gaston KJ (1992) Regional numbers of insect and plant species. Functional Ecology 6: 243-247

Gorman ML (1979) Island Ecology. Chapman and Hall, London pp 1-79

Gould SJ, Raup DM, Sepkoski JJ, Schopf TJM, Simberloff DS (1977) The shape of evolution: A comparison of real and random clades. Paleobiology 3: 23-40

Greuter W (1972) The relict element of the flora of Crete and its evolutionary significance. In Valentine DH [editor] "Taxonomy, phytogeography and evolutionary significance" Academic Press, London pp 161-177

Greuter W (1979) The origins and evolution of island floras as exemplified by the Aegean archipelago. In Bramwell D [editor] "Plants and Islands" Academic Press, London pp 87-106
Greuter W (1995) Extinctions in Mediterranean areas. In Lawton JH, May RM [editors] " Extinction Rates" Oxford University Press, Oxford pp 88-97
Grove SJ (2002) The influence of forest management history on the integrity of the saproxylic beetle fauna in an Australian lowland tropical rainforest. Biological Conservation 104: 149-171
Hammond PM (1990) Insect abundance and diversity in the Dumoga-Bone National Park, N. Silawesi, with special reference to the beetle fauna of lowland rain forest in the Toraur region. In "Insects and the Rain Forests of South East Asia (Wallacea). Knight WJ, Holloway JD (eds) Royal Entomological Society, London pp 197-254
Hammond P (1992) Species inventory. In B. Groombridge, ed. Global biodiversity: status of the earth's living resources. Chapman Hall, London. pp 17-39
Harvey PH (1995) Reply to Williams P. Trends in Ecology & Evolution 10: 82
Heywood VH, Mace GM, May RM, Stuart SN (1994) Uncertainties in extinction rates. Nature 368: 105
Hill JK, Thomas CD, Huntley B (1999) Climate and habitat availability determine 20th century changes in a butterfly's range margin. Proceedings of the Royal Society of London, Series B Biological Sciences 266: 1197-1206
Hill JK, Collingham YC, Thomas CD, Blakeley DS, Fox R, Moss D, Huntley B (2001) Impacts of landscape structure on butterfly range expansion. Ecology Letters 4: 313-321
Hodar JA, Castro J, Zamora R (2003) Pine processionary caterpillar *Thaumetopoea pityocampa* as a new threat for relict Mediterranean Scots pine forests under climatic warming. Biological Conservation 110: 123-129
Hodkinson ID, Casson D (1991) A lesser predilection for bugs: Hemiptera (Insecta) diversity in tropical forests. Biological Journal of the Linnean Society 43: 101-109
Hughes JB, Daily GC, Ehrlich PR (2000) Conservation of insect diversity: a habitat approach. Conservation Biology 14: 1788-1797
Idso SB, Idso CD, Idso KE 2003 The Specter of Species Extinction: Will Global Warming Decimate Earth's Biosphere? The George Marshall Institute Press pp 1-51
IUCN (World Conservation Union) (2002) Red list of threatened species. Available from http://www.redlist.org. IUCN, Gland, Switzerland
Jablonski D (2001) Lessons from the past: Evolutionary aspects of mass extinctions. Proceedings of the National Academy of Sciences 98: 5393-5398
Jablonski D (2002) Survival without recovery from mass extinctions. Proceedings of the National Academy of Sciences 99: 8139-8144
Jensen MN (2004) Climate warming shakes up species. BioScience 54: 722-729
Jonsson BG, Kruys N, Ranius T (2005) Ecology of species living on dead wood - lessons from dead wood management. Silva Fennica 39: 289-309

Kazakis G, Ghosn D, Vogiatzakis IN, Papanastasis VP (2006) Vascular plant diversity and climate change in the alpine zone of the Lefka Ori, Crete. Biodiversity and Conservation in press

Kim KC (1993a) Biodiversity, conservation and inventory: why insects matter? Biodiversity and Conservation 2: 191-214

Kim KC (1993b) Global biodiversity and conservation of insects. Biodiversity and Conservation 2: 189-190

Koh LP, Dunn RR, Sodhi NS, Colwell RK, Proctor HC, Smith VS (2004) Species coextinctions and the biodiversity crisis. Science 305:1632-34

Krebs CJ (1983) Ecology, The Experimental Analysis of Distribution and Abundance. Harper and Row, New York 678

Labandeira CC (2005) The fossil record of insect extinction: new approaches and future directions. American Entomologist 51: 1429

Labandeira CC, Phillips TL (1996) Insect feeding on Upper Pennsylvanian tree ferns (Palaeodictyoptera, Marattiales) and the early history of the piercing-and-sucking functional feeding group. Annals of the Entomological Society of America 89: 157-183

Labandeira CC, Sepkoski JJ (1993) Insect diversity in the fossil record. Science 261: 310-315

Lawton JH, May RM (1995) [editors] Extinction Rates. Oxford University Press, Oxford pp 1-233

Lawton JH, Schroeder D (1977) Effects of plant type, size of geographical range and taxonomic isolation on number of insect species associated with British plants. Nature 265:317-

Legakis A, Kypriotakis Z (1994) A biogeographical analysis of the island of Crete, Greece. Journal of Biogeography 21: 441-445

Londre RA, Schnitzer SA (2006) The distribution of lianas and their change in abundance in temperate forests over the past 45 years. Ecology 87: 2973-2978

Lundgren JG, (Organizer) (2005) Ground beetle (Coleoptera: Carabidae) ecology: their function and diversity in natural and agricultural habitats. American Entomologist 51: 218-239

MacArthur RH Wilson EO (1967) The Theory of Island Biogeography. Princeton University Press, Princeton, USA.

Mawdsley NA, Stork NE (1995) Species extinctions in insects: ecological and biogeographical considerations. In Harrington R, Stork NE [editors] "Insects in a changing environment" Academic Press, London pp 321-369

Maxwell WD (1989) The End Permian Mass Extinction, Mass Extinctions: Processes and evidence, Donovan, SK, Ed, London: Belkhaven, pp 152–173.

May RM (1975) Patterns of species abundance and diversity. In Codyy ML, Diamond JM Belknap Press, Harvard University Press, Cambridge, Massachusetts pp 81-120

May RM (1988) How many species are there on earth? Science 241: 1441-1449

May RM (1990) Taxonomy as destiny. Nature 347: 129-130

May RM (1992) How many species inhabit the earth? Scientific American 267: 42-48

McKinney ML (1995) Extinction selectivity among lower taxa: gradational patterns and rarefaction error in extinction estimates. Paleobiology 21: 300-313

McKinney ML (1997) Extinction vulnerability and selectivity: combining ecological and palaeontological views. Annual Review of Ecology and Systematics 28: 495-516

McKinney ML (1999) High rates of extinction and threat in poorly studied taxa. Conservation Biology 13: 1273-1281

McKinney ML (2002) Do human activities raise species richness? Contrasting patterns in United States plants and fishes. Global Ecology & Biogeography 11: 343-348

Menendez R, Gonzalez-Megias A, Collingham Y, Fox R, Roy DB, Ohlemueller R, Thomas CD (2007) Direct and indirect effects of climate and habitat factors on butterfly diversity. Ecology 88: 605-611

Mita E, Tsitsimpikou C, Tziveleka L, Petrakis PV, Ortiz A, Vagias K, Roussis V (2002). Seasonal variation of oleoresin terpenoids from *Pinus halepensis* and *P. pinea* and host selection of the scale insect *Marchalina hellenica* (Gennadius) (Homoptera, Coccoidea, Margarodidae, Coelostonidiinae). Holzforschung 56: 572-578

Mitrakos K (1980) A theory for Mediterranean plant life. Acta Oecologica / Oecologia Plantarum 1: 245-252

Moritz C, Faith DP (1998) Comparative phylogeography and the identification of genetically divergent areas for conservation. Molecular Ecology 7: 419-429

Morris MG (1987) Changing attitudes to nature conservation: the entomological perspective. Biological Journal of the Linnean Society 32: 213-323

Morris MG, Collins NM, Vane-Wright RI, Waage J (1991) The utilization and value of non-domesticated insects. In "Conservation of Insects" Collins NM, Thomas JA (eds) London: Academic Press pp 319-347

Morrone JJ (1994) On the identification of areas of endemism. Systematic Biology 43: 438-441

Morrow PA, LaMarche Jr VC (1978) Tree ring evidence for chronic insect suppression of productivity in subalpine Eucalyptus. Science 201: 1244-1246

Munguira MI, Thomas JA (1992) Use of road verges by butterfly and burnet populations and the effect of roads on adult dispersal and mortality. Journal of Animal Ecology 29: 316-329

Myers N (1993) Question of mass extinction. Biodiversity and Conservation 2: 2-17

Novotny V, Miller SE, Cizek L, Leps J, Janda M, Basset Y, Weiblen GD, Darrow K (2003) Colonizing aliens: caterpillars (Lepidoptera) feeding on Piper aducum and P. umbellatum in rainforests of Papua New Guinea. Ecological Entomology 28: 704-716

Panzer R, Schwartz MW (1998) Effectiveness of a vegetation based approach to insect conservation. Conservation Biology 12: 693-702

Parmesan C (1996) Climate and species' range Nature 382: 765-766

Parmesan C, Ryrholm N, Stefanescu C, Hill JK, Thomas CD, Descimon H, Huntley B, Kaila L, Kullberg J, Tammaru T, Tennet WJ, Thomas JA, Warren

M (1999) Polarward shifts in geographical ranges of butterfly species associated with regional warming. Nature 399: 579–583

Penuelas J, Estiarte M (1998) Can elevated CO2 affect secondary metabolism and ecosystem function? Trends in Ecology & Evolution 13: 20-24

Petrakis PV (1989) A multivariate approach to the analysis of biotopes structure with special reference to their avifauna in Prespa region, northwestern Greece. 4th Conference on Biogeography Ecology of Greece.

Petrakis PV (1991) Plant - Heteroptera Associations in an East -Mediterranean Ecosystem: An analysis of Structure, Specificity and Dynamics. PhD Dissertation University College of Cardiff, Wales, UK 0,

Petrakis PV (1992) Species area relationship derived from bird species on northern Sporades island complex: An attempt for dissection and reasoning. Biologia Gallo-hellenica 19: 129-142

Petrakis PV, Legakis A (2005) Insect migration and dispersal with emphasis on Mediterranean ecosystems. In Elewa AMT [editor] Migration of Organisms: Climate, Geography, Ecology: Causes of Migration in Organisms Springer Verlag, Berlin pp 85-126

Petrakis PV, Park S-J (2007) Hypericum Species in Agriculture and Pharmaceutical: Important Genotypes and Mechanism of Action. Hellenic Secretariat of Science & Technology pp 1-80

Petrakis PV, Roussis V, Ortiz A (2000) Geographic monoterpenoid variation in relation to morphology of Pinus brutia and Pinus halepensis in an east Mediterranean area (Attiki, Greece): implications for pine evolution. Edinburgh Journal of Botany 57: 349-375

Petrakis PV, Tsoukatou M, Vagias C, Roussis V, Cheng L (2003) Evolution probing for semiochemicals based on secondary metabolites in the cuticles of three species of Halobates (Heteroptera: Gerridae). Biological Journal of the Linnean Society 80: 671-688

Petrakis PV, Couladis M, Roussis V (2004) A method for detecting the biosystematic significance of the essential oil composition of five Hellenic Hypericum L. species. Biochemical Systematics and Ecology 33: 873-898

Pimm SL and Raven P (2000) Extinction by numbers. Nature 403: 843-845

Pimm SL, Moulton MP, Justice LJ (1995) Bird extinctions in the central Pacific. In Lawton JH, May RM (eds) Extinction Rates. Oxford University Press, Oxford, 75-87

Posadas P, Esquivel DRM, Crisci JV (2001) Using phylogenetic diversity measures to set priorities in conservation: an example from southern South America. Conservation Biology 15: 1325-1334

Purvis A, Jones KE, Mace GM (2000) Extinction. BioEssays 22: 1123-1133

Ranius T (2002) Osmoderma eremita as an indicator of species richness of beetles in tree hollows. Biodiversity and Conservation 11: 931-941

Ranius T, Aguado LO, Antonsson K, Audisio P, Ballerio A, Carpaneto GM, Chobot K, Gjurašin B, Hanssen O, Huijbregts H, Lakatos F, Martin O, Neculiseanu Z, Nikitsky NB, Paill W, Pirnat A, Rizun V, Ruiconescu A, Stegner J, Süda I, Szwalko P, Tamutis V, Telnov D, Tsinkevich V, Versteirt V, Vignon V, Vögeli M, Zach P (2005) Osmoderma eremita (Coleoptera,

Scarabaeidae, Cetoniinae) in Europe. Animal Biodiversity and Conservation 28: 1-44

Ranius T, Kindvall O (2006) Extinction risk of wood-living model species in forest landscapes as related to forest history and conservation strategy. Landscape Ecology 21; 687-698

Raup DM (1991) A kill curve for Phanerozoic marine species. Paleobiology 17: 37-46

Raup DM (1996) Extinction models. In Jablonski D, Erwin DH, Lipps JH (editors) "Evolutionary Paleobiology" The University of Chicago Press pp 419-433

Rhymer JM, Simberloff D (1996) Extinction by hybridization and introgression. Annual Review of Ecology and Systematics 27: 83-109

Ribera I, Vogler AP (2004) Speciation of Iberian diving beetles in Pleistocene refugia (Coleoptera, Dytiscidae). Molecular Ecology 13: 179-193

Ricklefs RE, Birmingham E (2002) The concept of the taxon cycle in biogeography. Global Ecology, Biogeography 11: 353-361

Ricklefs RE, Cox GW (1972) Taxon cycles in the West Indian avifauna. American Naturalist 106: 195-219

Ricklefs RE, Cox GW (1978) Stage of taxon cycle, habitat distribution and population density in the avifauna of the west Indies in the Wes Indian avifauna. American Naturalist 112: 353-361

Riemer J, Whittaker JB (1989) Air pollution and insect herbivores: Observed interactions and possible mechanisms. In Bernays EA (editor) Insect Plant Iteractions Volume I. CRC Press, Boca Raton

Ries L, Debinski DM, Wieland ML (2001) Conservation value of roadsideprairie restoration to butterfly communities. Conservation Biology 15: 401-411

Root TL, Price JT, Hall KR, Schneider SH, Rosenzeig C, Pounds JA (2003) Fingerprints of global warming on wild animals and plants. Nature 421: 57-60

Roslin T (2001) Large-scale spatial ecology of dung beetles. Ecography 24: 511-524

Ruf C, Fiedler K (2000) Thermal gains through collective metabolic heat production in social caterpilalrs of Eriogaster lanestris. Natirwssenschaften 87: 193-196

Saarinen K, Valtonen A, Jantunen J, Saarnio S (2005) Butterflies and diurnal moths along road verges: Does road type affect diversity and abundance? Auteur(s) / Author(s) Biological Conservation 123: 403-421

Samways ML (1996) Insects on the brink of major discontinuity. Biodiversity and Conservation 5: 1047-1058

Samways MJ (2006) Insect extinctions and insect survival. Conservation Biology 20: 245-246

Sanderson MJ (1997) A nonparametric approach to estimating divergence times in the absence of rate constancy. Molecular Biology and Evolution 14: 1218-1231

Schmitt T (2003) Biogeography and ecology of southern Portuguese butterflies and burnets (Lepidoptera). In Reemer M, van Helsdingen PJ, Kleukers RMJC (eds) Changes in ranges: invertebrates on the move. Proceedings of the 13th Inernational Colloquium of the European Invertebrate Survey, Leiden, 2-5 September 2001. EIS-Nederland, Leiden 69-78

Scott D, Malcolm JR, Lemieux C (2002) Climate change and modelled biome representation in Canada' s national park system planning and park mandates. Global Ecology & Biogeography 11: 475-484

Seabloom EW, Dobson AP, Stoms DM (2002) Extinction rates under nonrandom patterns of habitat loss. Proceedings of the National Academy of Sciences 99: 11229-34

Shaw P (2005) Estimating local extinction rates over successive time frames. Biological Conservation 121: 281-287

Siemann ED, Tilman D, Haarstad J, Ritchie M (1998) Experimental tests of the dependence of arthropod diversity on plant diversity. American Naturalist 152: 738-750

Signor PW, Lipps JH (1992) Sampling bias, gradual extinction patterns, and catastrophes in the fossil record. Geological Society of America Special Papers 190: 291-296

Small EC, Sadler JP, Telfer MG (2003) Carabid beetle assemblages on urban derelict sites in Birmingham. UK Journal of Insect Conservation 6: 233-246

Spies TA, Hemstrom MA, Youngblood A, Hummel S (2006) Conserving Old-Growth Forest Diversity in Disturbance-Prone Landscapes. Conservation Biology 20: 351-362

Steffan-Dewenter I, Tscharntke T (2002) Insect communities and biotic interactions on fragmented calcareous grasslands-a mini review. Biological Conservation 104: 275-284

Stork NE (1993) How many species are there? Biodiversity and Conservation 2: 215-232

Stork NE, Lyal CHC (1993) Extinction or 'co-extinction' rates. Nature 366; 307

Strong DR, Lawton JH, Southwood R (1984) Insects on plants: community patterns and mechanisms. Blackwell,Oxford,England.

Swofford DL (2002) PAUP, Phylogenetic Analysis Using Parsimony and Other Methods. Version 4.0 Sinauer Associates, Sunderland, Massachusetts

Tallamy DW (2004) Do alien plants reduce insect biomass? Conservation Biology 18: 1689-1692

Tewksbury L, Casagrande R, Bloosey B, Hafliger P, Schwarzlaender M (2002) Potential for biological control of *Phragmites australis* in North America. Biological Control 23: 628-630

Thanos CA, Fournaraki C, Georghiou K, Dimopoulos P, Bergmeir E (2005) A pilot network of plant micro-reserves in western Crete In "Book of Abstracts" XVII International Botanical Congress, Vienna, Austria pp 598

Thomas CB, Cameron A, Green RE, Bakkenes M, Beaumont LJ, Collingham YC, Erasmus BFN, Ferreira de Siqueira M, Grainger A, Hannah L, Hughes L, Huntley B, van Jaarsveld AS, Midgley GF, Miles L, Ortega-Huerta MA, Townsend Peterson A, Philips OL, Williams P (2004) Extinction risk from climate change. Nature 427: 146-148

Tutin TG, Heywood VH, Burges NA, Moore DM, Valentine DH, Walters SM, Webb DA (1964-1980) Flora Europaea. 5 Volumes, Cambridge University Press, Cambridge

Trabaud L (1991) Is fire an agent favouring plant invasions? Biogeography of Mediterranean Invasions Groves RH, Di Castri F Cambridge University Press, Cambridge, UK 0, 179-190

Tscharntke T, Kruess A (1999) Habitat fragmentation and biological control. In BA awikins and HV Cornell. Theoretical Approaches to Biological Control Cambridge University Press, Cambridge, UK pp 190-205

Tscharntke T, Steffan-Dewenter, I, Kruess A, Thies C (2002) The contribution of small habitat fragments to the conservation of insect communities of grassland-cropland landscape mosaics. Ecological Applications 12: 354-363

Valentine DH (1972) Taxonomy, Phytogeography and Evolution. Academic Press, London, 1-399

Van Valen LM (1984a) A resetting of Phanerozoic community evolution. Nature 307: 50-52

Van Valen LM (1984b) Strategies, Causes, and Aristotle. BioScience 34: 602

Vane-Wright RI, Humphries CJ, Williams PH (1991) What to protect? Systematics and the agony of choice. Biological Conservation 55: 235-254

Whalley P (1987) Insects and Cretaceous mass extinction. Nature 327: 562

Williams P (1995) The WORLDMAP debate. Trends in Ecology, Evolution 10: 82

Williams PH (2000) WORLDMAP in WINDOWS: Software and help document version 4.2 Privately distributed, London

Wilson EO (1961) The nature of the taxon cycle in the Melanesian ant fauna. American Naturalist 95: 169-193

Wilson EO (1987) The little things that run the world: the importance and conservation of invertebrates. Conservation Biology 1: 344-346

Wolters V, Bengtsson J, Zaitsev AS (2006) Relationship among the species richness of different taxa. Ecology 87:1886-1895

Yodzis P (1989) Introduction to Theoretical Ecology Harper and Row, New York pp 384

Zschokke S, Dolt S, Rusteholz H-P, Oggier P, Brascler B, Thommen GH, Ludin E, Erhardt A, Baur B (2000) Short-term responses of plants and invertebrates to experimental small-scale fragmentation. Oecologia 125: 559-572

Zurlini G, Grossi L, Rossi O (2002) Spatial-accumulation pattern and extinction rates of Mediterranean flora as related to species confinement to habitats in preserves and larger areas. Conservation Biology 16: 948-963

Index

acanthodians, 61
Africa, 162
ammonites, 69, 106
archaeological evidence, 162
Asian gazelle, 173

biomass, 163
bivalves, 73
blastoids, 61
bolide, 149
bony fish, 61
brachiopods, 61
bryozoans, 88

carbon content, 140
cataclysmic event, 129
causes, 149
Cenomanian/Turonian, 103
Central Sinai, 155
climate change, 163
Cluster analysis, 135
Co-extinctions, 223
conodonts, 78
conservation, 227
Cretaceous, 1

crinoids, 61, 88
Crurotarsi, 63
Current mass extinction, 191

debate, 160
decomposers, 205
dendrogram, 151
Devonian, 1
Devonian Plant Hypothesis, 60
Dinosaurs, 129
Dubois's antelope, 172
dwarfing, 161

echinoderms, 61
echinoids, 88
ecological generalists, 153
ecological specialists, 153
Egypt, 124
eurypterids, 61
extreme volcanism, 61

filterfeeders, 199
flourishment, 191
foraminifera, 61
Frasnian-Famennian boundary, 59
frozen methane hydrate, 61

Gastropoda, 108
Geographic Information Systems, 17
giant ape, 173
giant hyena, 174
giant panda, 176
giant tapir, 176
glaciation, 12
global warming, 207
greenhouse effect, 61

Humanity, 191
Hybridization, 225

ice age, 9
impact event, 61
Insects, 195
Inter-Tropical Convergence Zone, 168
introgression, 225
isotope analysis, 138

Javan rhinoceros, 174

K-Pg, 129

macrofaunas, 103
macroflora, 79
macroinvertebrates, 107
Malayan tapir, 175
Mass extinction, 1
media, 194
megafauna, 160
microflora, 79
modern extinction crisis, 181

natural protections, 194
neighbor joining algorithm, 151
North African Plate, 133
Northern Sinai, 155

orang-utan, 177
Ordovician, 1
ostracodes, 61

oxygen excursion, 139
oysters, 112

Paleogene, 155
Patterns, 149
peace, 191
pelycosaurs, 61
Permian, 1
Permian-Triassic, 61
pig, 174
placoderms, 61
Planktonic foraminifera, 149
plate tectonics, 61
Pleistocene, 162
prehistoric hunters, 161
principal coordinate analysis, 137

qualitative results, 135
quantitative results, 135
Quaternary, 159

Radiolaria, 71
reef builders, 76
robust macaque, 173
rugose and tabulate corals, 61

serow, 176
sharks, 61
Southeast Asia, 163
spotted hyena, 176
stegodons, 175
supernova, 61

tetrapod, 81
therapsids, 63
Triassic, 1, 65

UNEP, 194
UPGMA, 153

vascoceratides, 112
Vegetation, 168

Yunnan horse, 172

DATE DUE

DEMCO, INC. 38-2931